DESIGN OF CONCRETE STRUCTURE AND SOFTWARE APPLICATION

混凝土结构设计及软件应用

主编
苏 波
(Su Bo)

副主编
郝键铭 韩向科 于国军

主审
钱若军

参编
操礼林 刘 洋 陈家乐
张 抟 唐诗思 陈 智
Daniel Kumah 尹丹宁

镇江

图书在版编目(CIP)数据

混凝土结构设计及软件应用=Design of Concrete Structure and Software Application：英文/苏波主编. —镇江：江苏大学出版社,2019.12
ISBN 978-7-5684-1026-7

Ⅰ.①混… Ⅱ.①苏… Ⅲ.①混凝土结构－结构设计－计算机辅助设计－应用软件－高等学校－教材－英文 Ⅳ.①TU370.4-39

中国版本图书馆 CIP 数据核字(2019)第 154413 号

混凝土结构设计及软件应用
Design of Concrete Structure and Software Application

主　　编/	苏　波
责任编辑/	吴蒙蒙
出版发行/	江苏大学出版社
地　　址/	江苏省镇江市梦溪园巷 30 号(邮编：212003)
电　　话/	0511-84446464(传真)
网　　址/	http://press.ujs.edu.cn
排　　版/	镇江文苑制版印刷有限责任公司
印　　刷/	虎彩印艺股份有限公司
开　　本/	787 mm×1 092 mm　1/16
印　　张/	12
字　　数/	422 千字
版　　次/	2019 年 12 月第 1 版　2019 年 12 月第 1 次印刷
书　　号/	ISBN 978-7-5684-1026-7
定　　价/	45.00 元

如有印装质量问题请与本社营销部联系(电话：0511-84440882)

Preface

More and more overseas students come to China now. Each year, there are more than 100 students enrolling in Jiangsu University and about 20 students of them major in Civil Engineering. From the year of 2012, the chief editor started to teach the course "Design of Concrete Structure" for oversea students majoring in Civil Engineering. The course has been listed one of "Excellent Courses Taught by English for International Students in Jiangsu University" from 2016, but there is still no proper textbook suitable for overseas students. To meet the requirement of teaching, from 2018, the editor started to compile this book based on the teaching experience for overseas students.

This book mainly focuses on practical teaching and aims at cultivating students' basic skills, comprehensive application of process knowledge, engineering quality and innovation consciousness. It is divided into 5 chapters, the main contents include: Principles of structural Design, Slabs, Single-Storey Factory Building, Multilayer Frame Structures, Application of Software.

Considering the characteristics of overseas students, we expected to achieve the following three special objectives:

(1) Offer the overseas students a reference English textbook for concrete structure design.

(2) Simplify some contents with abstract theory, and give more practical examples.

(3) Application of software ETABS is elaborated in the textbook to meet the requirement of international students' final project.

The chief editor of this book is Su Bo. The associate editor is Hao Jianming, Han Xiangke and Yu Guojun. The reviewer is Qian Ruojun. There are many students have participated in the compiling of the book, such as Chen Jiale, Zhan Tuan, Tang Shisi, Chen Zhi, Jiang Wei, Wang Yue, Liu Yaling, Zhang Qingqing, Liu Peng, Daniel Kumah, Yin Danning, and so on. I must say thanks for their hard work during the process of making draft, translation, drawing and modification. Also many colleagues have given me a lot of help during the edition, such as Cao Lilin, Li Xinchao, Liu Yang, Zhang Xiuli, Yang Fan, and so on.

A large amount of relevant teaching material, manual and literature at home and abroad was consulted and cited during compiling, I would like to pay my respects and thanks to the authors concerned.

Due to the limited level of editors, this textbook inevitably might have some errors, peer and readers rectification is accepted!

Su Bo

2019.10.1

Contents

Chapter 1 Principles of Structural Design / 001

1.1 Introduction / 001
    1.1.1 Structural Forms / 001
    1.1.2 Concrete Type / 002
    1.1.3 Objectives of Building Structural Design / 003
    1.1.4 Structural System / 004
1.2 Loads / 005
    1.2.1 Loads Categories / 005
    1.2.2 Representative Values of Loads / 005
    1.2.3 Permanent Load / 006
    1.2.4 Live Loads / 007
    1.2.5 Wind Load / 009
1.3 Material / 013
    1.3.1 Concrete / 013
    1.3.2 Steel Reinforcements / 014
1.4 Structural Design Theory / 014
    1.4.1 Limit States Design / 014
    1.4.2 Load Combinations / 015
1.5 The Design Process / 017
    1.5.1 Definis Client's Needs and Priorities / 017
    1.5.2 Develop Project Concept / 017
    1.5.3 Design Individual Systems / 017
References / 018
Questions / 018

Chapter 2 Slabs / 019

2.1 Introduction / 019
    2.1.1 Types of Floor Systems / 019
    2.1.2 Size of Reinforced Concrete Slabs / 022
2.2 Design of One-Way Ribbed Slab / 023
    2.2.1 Structure Layouts of One-Way Slab / 023

2.2.2 Calculating Diagram of One-Way Slab / 024
2.2.3 Internal Forces Analysis by Elastic Theory Method / 025
2.2.4 Internal Forces Analysis by Plastic Theory Method / 028
2.2.5 Section Design and Reinforcements Construction of One-Way Slabs / 032

2.3 Design of Two-Way Ribbed Slab / 037
2.3.1 Structure Layouts of Two-Way Ribbed Slab / 037
2.3.2 Stress Characteristics of Two-way Ribbed Slab / 037
2.3.3 Analysis of Two-Way Slab's Internal Forces by Elastic Theory Method / 037
2.3.4 Internal Forces Analysis of Two-Way Slab using the Plastic Theory Method / 039
2.3.5 Internal Force Analysis of Two-Way Slab Beam / 044
2.3.6 Section Design and Construction Requirement of Two-Way Slab / 045
2.3.7 Other Floor Types / 047
2.3.8 Platform Slab of a Workshop / 048

2.4 Stairs / 050
2.4.1 Slab Stairs / 051
2.4.2 Beam Stairs / 051

References / 052
Questions / 052

Chapter 3 Single-Storey Factory Building / 054

3.1 Introduction / 054
3.2 Structural System and Design of Single-Layer Factory Buildings / 055
3.2.1 Structure System / 055
3.2.2 Layout of Structure Members / 058
3.3 Internal Force Calculation of Bent Frame Structure / 064
3.3.1 Assumptions and Calculations / 064
3.3.2 Standard Load Values of Bent Frame Structure / 065
3.3.3 Internal Force Analysis / 071
3.3.4 Space Effect in Internal Force Analysis of Single-Storey Bent Frame Structure / 079
3.3.5 Internal Force Combination / 082
3.4 Bent Frame Column of Single-Storey Factory / 084
3.4.1 Column Form / 085

3.4.2 Design of Bent Frame Column / 088
3.5 Design of Crane Beam / 097
    3.5.1 Mechanical Characteristics of Crane Beam / 097
    3.5.2 Selection of Crane Beam / 098
    3.5.3 Internal Force Calculation of Crane Beam / 099
References / 100
Questions / 100

Chapter 4 Multilayer Frame Structures / 102

4.1 Frame Construction Types and Layout / 102
    4.1.1 Features and Types of Frame Structures / 102
    4.1.2 Structure Layouts / 102
    4.1.3 Structure Form Selection and Section Dimension Estimating / 104
4.2 Internal Force and Displacement Calculation / 105
    4.2.1 Analysis Models / 106
    4.2.2 Internal Force Analysis under Vertical Loads / 107
    4.2.3 Internal Force Analysis under Horizontal Loads / 110
    4.2.4 Calculation of Displacement and the Limit Value of Frames / 119
4.3 The Most Unfavourable Position of Load Combination / 123
    4.3.1 Control Sections / 123
    4.3.2 Load Effect Combination / 123
    4.3.3 Most Unfavorable Position under Vertical Variable Loads / 125
    4.3.4 Moment Amplitude Modulation of Beam End / 126
4.4 Design of Frame Structural Components / 127
    4.4.1 Design of Frame Beam and Column / 127
    4.4.2 Design of Frame Joints / 127
    4.4.3 Frame Construction Requirements / 131
References / 134
Questions / 134

Chapter 5 Application of Software / 136

5.1 Introduction of Softwares / 136
    5.1.1 Introduction of Chinese Softwares / 136
    5.1.2 Introduction of International Softwares / 138

5.2 Example: Design of a Reinforced Concrete Structure / 139
    5.2.1 Introduction of Project / 139
    5.2.2 Begin a New Model / 140
    5.2.3 Add Floor Openings / 145
    5.2.4 Add Walls / 148
    5.2.5 Define Static Load Patterns / 150
    5.2.6 Review Diaphragms / 151
    5.2.7 Review the Load Cases / 154
    5.2.8 Run the Analysis / 155
    5.2.9 Display the Results / 157
    5.2.10 Design the Concrete Frames / 159
    5.2.11 Design the Shear Walls / 163
References / 166

Appendix / 167

Appendix A / 167
Appendix B / 170
Appendix C / 173
Appendix D / 178

Chapter 1

Principles of Structural Design

1.1 Introduction

1.1.1 Structural Forms

According to the number of floors, building structures can be divided into single-storey, multi-storey, high-rise building and super high-rise buildings (Figure 1.1-1.3). The Chinese code, *Technical specification for concrete structures of tall building in China* (Ref. 1.1) stipulates that residential buildings which have 10 storeys or more, or the height of which is above 28m, or other civilian buildings of which the height is above 24m are specified as high-rise buildings. Generally, buildings which are above 100m high are called super high-rise buildings. Long-span structures are special buildings which are widely used for gymnasiums, exhibition halls, bridges and other large-span buildings which need advanced design and construction technologies (Figure 1.4, 1.5).

Based on type of materials used, buildings can be divided into wooden frames structures, masonry structures, concrete structures, steel structures, composite structures, and so on. Wood was once the main material of Chinese ancient buildings. However, for forest protection and fire prevention requirements, wood is rarely used now. Recently, glued laminated timber structure is on a slow rise. Due to poor tensile strength, masonry materials are not suitable for the horizontal members. Pure masonry structures are very rare and usually mixed with other materials. Masonry materials are mainly used in vertical members. Composite structures are structures whose members or sections are made by two or more materials. The past few decades have seen some outstanding advances in the use of composite materials in structural applications. There can be little doubt that, within engineering circles, composite structures have changed traditional design concepts and made possible an unparalleled range of new and exciting possibilities of viable materials for construction.

FIGURE 1.1　A single-layer factory

FIGURE 1.2　A multi-layer teaching building

FIGURE 1.3　High-rise buildings and skyscrapers in Pudong District of Shanghai

FIGURE 1.4　Runyang Bridge across Yangtze River

FIGURE 1.5　Zhengjiang Sport Centre

1.1.2　Concrete Type

Concrete is a stone-like material obtained by mixing a reasonable proportion of cement, sand, gravel (or other aggregate) and water to harden in forms of the shape and dimensions of the desired structure. Concrete is a universal building material which can be cast and made to fill forms or molds of almost any practical shape. Its high fire and weather resistance is an evident advantage. Most of the constituent materials, with the exception of cement and additives are usually available at low cost locally or at a small distance from the construction site. Its compressive strength, like that of natural stone, is high, which makes it suitable for members primarily subject to compression, such as columns and arches. On the other hand, natural stone is a relatively brittle material whose tensile strength is small compared with its compressive strength.

In the second half of the nineteenth century, it was found that the tensile strength limitation of concrete can be offset by the use of steel with its high tensile strength to reinforce concrete, chiefly in those places where its low tensile strength would limit the carrying capacity of the member. The reinforcement, usually round steel rods with appropriate surface deformations to provide adequate interlocking capacity between concrete and steel, and it's placed in the weak tensile area of the concrete. When completely surrounded by the hardened concrete mass, it forms an integral part of the member. The resulting combination of two materials, known as reinforced concrete, complement each other: they are relatively cheaper, it has good weather and fire resistance, good compressive strength, and excellent formability of concrete and the high tensile strength and much greater ductility and toughness of steel. It is this combination that allows the almost unlimited range of uses and possibilities of reinforced concrete in the construction of buildings, bridges, dams, tanks, reservoirs, and other structures.

A special way has been found, however, to use steel and concrete of very high strength in combination. This type of construction is known as prestressed concrete. The steel, in the form of wires, strands or bars, is embedded in the concrete under high tension that is held in equilibrium by compressive stresses in the concrete after hardening. Because of this pre-compression, the concrete in a flexural member will crack on the tension side at a much larger load than that is not properly pre-compressed. Prestressing greatly reduces both the deflections and the tensile cracks at ordinary loads in such structures, and thereby enables these high-strength materials to be used effectively. Prestressed concrete has been extended to a very significant extent in the range of spans of structural concrete and the types of structures for which it is suited (Ref. 1.2).

1.1.3 Objectives of Building Structural Design

Structural frame is the building's main "skeleton" which bears various loads during the construction and service period, including the deadweight of the building, the gravity of people and furniture, snow and wind loads, earthquake effects, and so on. Considering the above loads and effects with a quantitative description that may arise in the future, Building Structural Design is to design a structure which is safe enough, meet the operating requirements and have a long service life, using mechanics and structural knowledge. Reasonable structural design must guarantee the building safety, serviceability and durability.

Safety requires that the strength of the structure is adequate for all loads that may foreseeable act on it. If the strength of a structure built as designed could be predicted accurately, and if the loads and their internal effects (moments, shears, axial forces) are known accurately, safety could be ensured by providing a carrying capacity just barely in excess of the known loads.

Serviceability requires the structural deflections be minimal; that cracks, if any, be kept to tolerable limits; that vibrations be minimized; etc.

Durability refers to the ability of a structure which completes the intended function without extensive repairs in a given working environment, within the intended service life, under normal maintenance conditions. Generally speaking, for concrete structure, durability includes impermeability, frost resistance, erosion resistance and other properties.

1.1.4 Structural System

For structural design, particular attention must be given to the structural system of the buildings. The Structural system is the structural elements composition which resists external actions as a system. Generally, a structural system is composed of horizontal, vertical systems and foundation.

The horizontal system consists of floor and roof structures. The main types of roof structure include slab-beam system, truss systems, grid system, arch shell system, cable and membrane system. Reinforced concrete floor structure can be divided into two categories: slab-beam and flat plate floor slab. Plate-beam floors are usually supported by bidirectional beams as shown in Figure 1.6, and can be dived into one-way slab, two-way slab and multi-ribbed slab. We will discuss them in detail in Chapter 2. Flat plate floor is used when a large indoor clearance without no beams is required, and it usually has column caps to improve shearing force, punching shear resistance capacity as shown in Figure 1.7.

FIGURE 1.6 Plate-beam floor

FIGURE 1.7 Flat plate floor

The vertical system has three fundamental types: frame structure, wall structure and tube structure. Frame structure is made up of beams and columns. When the connections are hinged, it is called bent frame structure. When the connections are rigid, it is called rigid frame structure, or frame structure for short. Concrete single layer factories are often bent frame structures, which are not sensitive to the differential settlement of foundations. Frame structure is the most current form of multi-storey structure.

Vertical members in wall structures are usually concrete walls whose height of the section is much larger than the thickness and can bear vertical and horizontal loads. Tubular structures are space structures made up of several pins of a shear wall with ring-shaped in plan, including frame-tube structure, tube-in-tube and bundled tubes structure. They are mainly used for high-rise buildings.

Foundation system contains basement and building foundation (Deep or Shallow Foundation). Independent foundation under column, strip foundation, cross base, raft foundation, box basis are shallow foundations. Pile foundation, underground continuous wall, Sunk shaft foundation are deep foundations.

These various forms of basic structures can be combined to form a composite structure, along horizontal and vertical direction. The former refers to using two or more basic structures in the horizontal plane, such as the frame-shear wall structure or the net-shell structure. Whereas the latter refers to using different structures in the vertical direction, such as the upper shear wall and bottom frame structure. You can also use different materials in the structure such as concrete- and steel-bent frame structure. With the development of science and technology and new construction requirements of the building, new structures and structural materials will arise.

1.2 Loads

1.2.1 Loads Categories

Load on structure can be classified into the following three types (Ref. 1.3):

① Permanent load, such as self-weight of structure, soil pressure, prestressing force, etc.

② Variable load, such as live load on floors, live load on roofs, crane load, wind load, snow load, etc.

③ Accidental load, such as explosive force, collision, etc.

1.2.2 Representative Values of Loads

Representative value of a load is the measuring value adopted for checking calculations of the limit states in design, such as the characteristic value, the combination value, the frequent value and the quasi- permanent value (Ref. 1.3).

Characteristic value: It is the fundamental representative value of a load, which denotes the characteristic value for the statistical distribution of the maximum load in the design reference period (such as mean-value, mid-value or certain fractile). The design reference period is selected for determining the representative values of the variable load and takes the values of 50 years according to Chinese code.

Combination value: For the values of variable loads after combination, that their transcendental probability for the load effects in the design reference period can be tended toward identical with the corresponding probability for the load effect of the appearance of single load alone; or the values of variable loads after combination, that the structure should has the unified stipulated reliability index.

Frequent value: For the value of variable load in the design reference period, that the transcendental total time is in the small ratio of stipulated time, or the transcendental frequency is the stipulated frequency.

Quasi permanent value: For the value of a variable load in the design reference period, that the transcendental total time is about one-half of the design reference period.

Different representative values shall be adopted for different loads in the design of building structures.

The characteristic value shall be adopted as the representative value of the permanent load. The characteristic value, combination value, frequent value or quasi-permanent value shall be adopted as the representative value of the variable load in accordance with the requirements of design. The representative value of accidental load shall be determined in accordance with the features and function of the building structures.

1.2.3 Permanent Load

In the service period of the structure, the value of the load is not varied with time, or the variation of load, which compared with a mean value can be neglected, or the variation of the load is in one sense and can attain some limiting value. Dead loads are those that are constant in magnitude and fixed in a particular location throughout the lifetime of the structure. Usually the major part of the dead load is the weight of the structure itself, such as the self-weight of roof, floor, wall, other components like beam and column, a heat insulating layer, waterproof layer, decorative material layer and the fixed equipment. Standard value of dead load can be calculated out by the design size of members and the material density which can be found in Chinese code (Ref. 1.3).

EXAMPLE 1.1

The floor beam in Fig. 1.8 is used to support the 2000mm width of a lightweight plain concrete slab having a thickness of 100mm. The slab serves as a portion of the ceiling of the floor below, and therefore its bottom is coated with plaster. Furthermore, a 2400mm high and 300mm thick lightweight solid concrete block wall is directly over the top flange of the beam. Determine the loading on the beam measured per meter length of the beam.

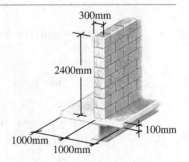

FIGURE 1.8 Part of a slab with block wall

SOLUTION

Concrete slab: 25 × 2 × 0.1 = 5kN/m

Plaster: 17 × 0.02 × 2 = 0.68kN/m

Block wall: 15 × 2.4 × 0.3 = 10.8kN/m

Total load: 5 + 0.68 + 10.8 = 16.48kN/m

1.2.4 Live Loads

Variable loads mean that in the service period of structure, the value of load is varied with time, or the variation of load, which compared with the mean value cannot be neglected. Live Load is one of the basic types of variable loads.

1.2.4.1 Live loads on floors

The characteristic value and the coefficients for combination value, frequent value and quasi-permanent value of uniform live loads on floors in civil buildings should be taken according to the stipulations in the Table 1.1 (Ref. 1.3).

TABLE 1.1 The characteristic value and other representative values of uniform live loads on floors in civil buildings

Item No.	Type	Chatacteristic value/(kN·m^{-2})	Coefficient for combination value ψ_c	Coefficient for frequent value ψ_f	Coefficient for quasi-permanent value ψ_q
1	(1) Dwelling, hostel, hotel, office, hospital ward, nursery, kindergarten (2) Laboratory, reading room, meeting room, hospital outpatient room	2.0	0.7	0.5 0.6	0.4 0.5
2	Classroom, canteen, dining-hall, ordinary archives	2.5	0.7	0.6	0.5
3	(1) Assembly hall, theater, cinema, grand-stands with fixed seats (2) Laundry	3.0 3.0	0.7 0.7	0.5 0.6	0.3 0.5
4	(1) Stores and shopes, exhibition hall, station, port, airport hall and waiting room for passengers (2) Stands without fixed seat	3.5 3.5	0.7 0.7	0.6 0.5	0.5 0.3
5	(1) Gymnasium, arena (2) Athletic field and dance hall	4.0 4.0	0.7 0.7	0.6 0.6	0.5 0.3
6	(1) Storehouse for collecting books, archives, storeroom (2) Densely bookcase storehouse	5.0 12.0	0.9 0.9	0.9 0.9	0.8 0.8
7	Ventilator motor room, elevator motor room	7.0	0.9	0.9	0.8

Continued

Item No.	Type	Chatacteristic value/(kN·m^{-2})	Coefficient for combination value ψ_c	Coefficient for frequent value ψ_f	Coefficient for quasi-permanent value ψ_q
8	Automobile passage and garage: (1) With one-way slab floor, (span of slab is not less than 2m) and two-way slab floor(span of slab is not less than 3m×3m)passenger car fire truck (2) with two-way slab (span of slab is not less than 6m×6m)and flat slab floor(column net work size is not less than 6m×6m) passenger train fire engine	4.0 35.0 2.5 20.0	0.7 0.7 0.7 0.7	0.7 0.5 0.7 0.5	0.6 0.0 0.6 0.0
9	Kitchen: (1) for dining hall (2) ordinary	2.0 4.0	0.7 0.7	0.7 0.6	0.7 0.5
10	Bathroom, toilet and washroom	2.5	0.7	0.6	0.5
11	Passage, entrance hall, staircase: (1) Hostel, hotel, nursery, hospital ward, kindergartens, dwelling-house (2) Office, classroom, dining hall, hospital outpatient department (3) Fire dispersed staircase, other civil buildings	2.0 2.5 3.5	0.7 0.7 0.7	0.5 0.6 0.5	0.4 0.5 0.3
12	Staircase: (1) multi-layered house (2) other	2.0 3.5	0.7 0.7	0.5 0.5	0.4 0.3
13	Balcony: (1) when population may be concentrated (2) other	3.5 2.5	0.7 0.7	0.6 0.6	0.5 0.5

A reduction coefficient of live load is used to consider the number of storeys in a building according to Table 1.2.

TABLE 1.2 Reduction coefficient of live load according to the number of storeys in a building

The number of storeys above the calculated section of walls, columns and foundations.	1	2~3	4~5	6~8	9~20	>20
Reduction coefficients of the total live loads on each floor above the calculated section.	1.0(0.90)	0.85	0.70	0.65	0.60	0.55

Note: The value in brackets is adopted when tributary area of the beam is larger than 25m^2.

1.2.4.2 Live loads on roofs

Uniform live loads that project on roofs shall be adopted according to Table 1.3.

TABLE 1.3 Uniform live loads on roofs

Item No.	Type	Chatacteristic value/(kN/m^2)	Coefficient for combination value ψ_c	Coefficient for frequent value ψ_f	Coefficient for quasi-permanent value ψ_q
1	Unmanned roof	0.5	0.7	0.5	0
2	Manned roof	2.0	0.7	0.5	0.4
3	Roof garden	3.0	0.7	0.6	0.5
4	Roof sports ground	3.0	0.7	0.6	0.4

Note:

① The unmanned roof, when the construction or maintenance loads are heavy, that the actual conditions shall be adopted; the stipulations of the relevant design codes shall be adopted for different structures, it is right to increase or decrease $0.2kN/m^2$ for characteristic values.

② The manned roof, which takes account of other uses. Its corresponding live load on floors shall be adopted.

③ The ponding load on the floor, which is caused due to the impeded drainage blocked up and etc. The detailing measures for precaution shall be taken; if necessary, its live load on roofing shall be determined by the possible depth of ponding.

④ The live load for the roof garden is not including the self-weight of soil, stone, materials in the flower nursery and etc.

1.2.5 Wind Load

1.2.5.1 Characteristic value of wind load

Characteristic value of wind load vertical to building surfaces shall be calculated in accordance with the following equation:

(1) When in the design of principal load-bearing structures

$$W_k = \beta_z \mu_s \mu_z w_0 \tag{1-1}$$

where, W_k—characteristic value of wind load, kN/m^2;

β_z—dynamic effect factor of wind at a height of z;

μ_s—shape factor of wind load;

μ_z—exposure factor for wind pressure;

w_0—the reference wind pressure, kN/m^2.

(2) When in the design of fencing structures

$$W_k = \beta_{gz} \mu_s \mu_z w_0 \tag{1-2}$$

where, β_{gz} is gust factor at the height of z.

1.2.5.2 Reference wind pressure

The wind pressure is determined as:

$$w_0 = \frac{1}{2}\rho v_0^2 \tag{1-3}$$

where, v_0—the reference wind speed, m/s, which is defined as the mean wind speed in 10 minutes over a flat and open terrain at an elevation of 10 m with the mean return period of 50 years;

ρ—the density of air, t/m^3.

The reference wind pressures of some Chinese cities are given in the Appendix A. 1. For high-rise structures and the other structures sensitive to wind load, the value shall be raised appropriately, and shall be specified in the relevant design codes μ_z.

1.2.5.3 Exposure factor for wind pressure

Terrain roughness is introduced to consider the terrain' effect on wind pressure which can be classified in A, B, C and D four categories:

Category A denotes in shore sea surfaces, islands, sea shores, lake shores and deserts;

Category B denotes open fields, villages, forests, hills, sparsely-populated towns and city suburbs;

Category C denotes urban districts in densely-populated cities;

Category D denotes densely-populated cities with high building urban districts.

The exposure factor μ_z is determined on the basis of the categories of terrain roughness and the height above terrain or sea level as shown in Table 1. 4.

TABLE 1.4 Exposure factor μ_z for wind pressure

Height about terrain or sea level/m	Terrain roughness categories			
	A	B	C	D
5	1.17	1	0.74	0.62
10	1.38	1	0.74	0.62
15	1.52	1.14	0.74	0.62
20	1.63	1.25	0.84	0.62
30	1.8	1.42	1	0.62
40	1.92	1.56	1.13	0.73
50	2.03	1.67	1.25	0.84
60	2.12	1.77	1.35	0.93
70	2.2	1.86	1.45	1.02
80	2.27	1.95	1.54	1.11
90	2.34	2.02	1.62	1.19
100	2.4	2.09	1.7	1.27
150	2.64	2.38	2.03	1.61
200	2.83	2.61	2.3	1.92
250	2.99	2.8	2.54	2.19
300	3.12	2.97	2.75	2.45

Continued

Height about terrain or sea level/m	Terrain roughness categories			
	A	B	C	D
350	3.12	3.12	2.94	2.68
400	3.12	3.12	3.12	2.91
≥450	3.12	3.12	3.12	3.12

1.2.5.4 Shape factor of wind load

The shape factor of a building or a structural construction can adopt the stipulations in the code. The shape factor of three different roof are given in Table 1.5 as examples. If the shape is not shown in the code or lack of reference materials, the wind tunnel test shall be used to determine the shape factors especially for the important and complicated shaped buildings.

TABLE 1.5 Shape factor of wind load of three roof

Item No.	Types	Shapes and shape factors μ_s		
1	Enclosed double pitched roof on the ground	For medium values, calculated by interpolation	α	μ_s
			0°	0
			30°	+0.2
			≥60°	+0.8
2	Enclosed double pitched roof	For medium values, calculated by interpolation	α	μ_s
			≤15°	−0.6
			30°	0
			≥60°	+0.8
3	Enclosed arched roof on the ground	For medium values, calculated by interpolation	F/L	μ_s
			0.1	+0.1
			0.2	+0.2
			0.5	+0.6

1.2.5.5 Dynamic wind effect factor

For the engineering structures, such as buildings, roofings and various high-rise buildings, with the fundamental natural period of vibration T_1 is greater than 0.25s, as well as, the high-flexible buildings with height is greater than 30m and the height-width ratio is greater than 1.5, that the influence of the along wind direction excitation due to the fluctuation effects of wind press shall be considered. The random vibration theory shall be used for calculation of the wind excitation, and the natural period of vibration for structure shall be calculated according to the dynamics of structures.

For the common cantilever-type structures, such as high-rise structures including the structural framing, tower frame, chimney and etc., as well as, the high buildings with the height is greater than 30m, the height-width radio is greater than 1.5 and the influence of torsion can be neglected that only the influence of the first vibration mode may be considered and the wind load on such structures can be calculated through dynamic wind effect factors according to equation 1.1, and the dynamic wind effect β_z at the height z of structures can be calculated according to the following equation:

$$\beta_z = 1 + \frac{\zeta v \varphi_z}{\mu_z} \tag{1-4}$$

where, ζ—magnification factor of wind fluctuation;
v—wind fluctuation factor;
φ_z—vibration mode factor;
μ_z—exposure factor for wind pressure.

For the structure with round section, the checking of the cross wind direction wind excitation (vortex shedding) shall be carried out on the basis of the different conditions for Reynolds number. There is no extension of introduction of this part, the reader can read the reference books for further detail (Ref. 1.3).

For single-layer buildings or multilayer buildings, β_z usually equals one for simplification, which means the dynamic effect are neglected for these buildings.

EXAMPLE 1.2

The horizontal plan of shear wall structure is shown in Fig. 1.9, the chart has sectioned out the wind load and joint acting position, the ground roughness is B, the height of the building is 58m, the structure is situated in the quite sparse villages, the basic wind pressure $w_0 = 0.385 kN/m^2$, gust factor $\beta_z = 1.41$, the wind direction are in the chart and the arrow shows the direction.

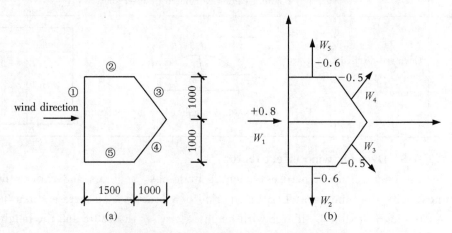

FIGURE 1.9 Wind load on shear wall structure

SOLUTION

(1) The ground roughness is B, the wind pressure height variation coefficient of 58m place is:

$\mu_z = 1.62 + \dfrac{58-50}{60-50} \times (1.71 - 1.62) = 1.69$

$\mu_z = 1.67 + \dfrac{58-50}{60-50} \times (1.77 - 1.67) = 1.75$

(2) The computational process is listed in the following Table 1.6.

TABLE 1.6 The computational process

Face number	μ_{si}	$B_i \cos\alpha_i$	$\mu_{si} B_i \cos\alpha_i$
①	0.8	20	16
③	−0.5	10	−3.535
④	−0.5	10	−3.535
		$\sum \mu_{si} B_i \cos\alpha_i$	23.07

(3) According to formula 1.1.

$w_z = \beta_z \mu_z w_0 \sum \mu_{si} B_i \cos\alpha_i = 1.41 \times 1.75 \times 0.385 \times 23.07 = 21.92 \text{kN/m}^2$

1.3 Material

In this book, we focus on Reinforced Concrete structures, which the structures and component members are made of Reinforced Concrete. An understanding of the material characteristics and behavior under load is fundamental to know the performance of structural concrete, and for safe, economical, and serviceable design of concrete structures. A brief review is presented in this section about the materials used in Reinforced Concrete structures.

1.3.1 Concrete

Concrete strength grade shall be determined in accordance with the characteristic value of cube compressive strength obtained from standard tests on 150mm side long cube specimen, which is fabricated and cured for a period of 28 days under standard conditions, and has the compressive strength with a 95% degree of veliability.

According to Chinese Load Code (Ref. 1.4), Characteristic values of axial compression strength f_{ck} and axial tension strength f_{tk} of different concrete grade are shown in Table 1.7.

TABLE 1.7　Characteristic values of concrete strength　　　(N/mm²)

Type of strength	Concrete strength grade													
	C15	C20	C25	C30	C35	C40	C45	C50	C55	C60	C65	C70	C75	C80
f_{ck}	10.0	13.4	16.7	201.1	23.4	26.8	29.6	32.4	35.5	38.5	41.5	44.5	47.4	50.2
f_{tk}	1.27	1.54	1.78	2.01	2.20	2.39	2.51	2.64	2.74	2.85	2.93	2.99	3.05	3.22

Design values of Concrete Strength are listed in Table 1.8.

TABLE 1.8　Design values of concrete strength　　　(N/mm²)

Type of strength	Concrete strength grade													
	C15	C20	C25	C30	C35	C40	C45	C50	C55	C60	C65	C70	C75	C80
f_c	7.2	9.6	11.9	14.3	16.7	19.1	21.1	23.1	25.3	27.5	29.7	31.8	33.8	35.8
f_t	0.91	1.10	1.27	1.43	1.57	1.71	1.80	1.89	1.96	2.04	2.09	2.14	2.18	2.22

1.3.2　Steel Reinforcements

The guarantee rate for the characteristic value of strength for steel bars shall not be less than 95%. Characteristic value of yield strength f_{yk} and ultimate strength f_{stk} for hot-rolled steel bars are listed in Table 1.9. Design values of tensile and compressive strength are also listed behind (Ref. 1.4).

TABLE 1.9　Characteristic value and design values of strength for ordinary steel bars (N/mm²)

Grade	d/mm	f_{yk}	f_{stk}	f_y	f_y'
HPB300	6~22	235	420	370	270
HRB335 HRBF335	6~50	335	455	300	300
HRB400 HRBF400	6~50	400	540	300	360
HRB500 HRBF500	6~50	500	630	435	410

1.4　Structural Design Theory

1.4.1　Limit States Design

1.4.1.1　Ultimate limit states

For ultimate limit states, the structural members of building shall accord with the fundamental combination or accidental combination of load-effects and adopt the following design expression (Ref. 1.3, Ref. 1.5):

$$\gamma_0 S_d \leqslant R_d \tag{1-5}$$

where, γ_0—factor of importance. For structural members of safety class I or the design service year is 100 years or above, γ_0 shall not be less 1.1; For structural members of safety class II or the design service year is 50 years, γ_0 shall not be less than 1.0; For structural members of safety class III or the design service year is less than 5 years, γ_0 shall not be less than 0.9. In earthquake-resistant design, the factor of importance for structural members may not be considered.

S_d—design value for combination of load-effects for ultimate limit states, which is calculated in accordance with the current National Standards 'Load code for design of building structures' (GB 50009), and 'Code for seismic design of buildings' (GB 50011).

R_d—design value of loading-bearing capacity of structural members. In the earthquake-resistant design, the value of R_d shall be divided by the earthquake-resistant adjusted coefficient γ_{RE}.

1.4.1.2 Serviceability limit states

For the serviceability limit states, the characteristic combination, the frequent combination or the quasi-permanent combination of loads shall be adopted in accordance with the different requirements of design, and adopt the following design expression:

$$S \leqslant C \qquad (1\text{-}6)$$

where, S—the combination value of load-effects of serviceability limit states;

C—the limiting value of deformation, width of cracks and stress etc., when the structural members of building structures meet the stipulated requirements for serviceability.

1.4.2 Load Combinations

For the fundamental load combination, the design value S of the combination loads effect shall be determined by the most unfavorable value taking the following combination values:

(1) When be combination is controlled by the variable load effect:

$$S_d = \sum_{j=1}^{m} \gamma_{Gj} S_{Gjk} + \gamma_{Q1} \gamma_{L1} S_{Q1k} + \sum_{i=1}^{n} \gamma_{Qi} \gamma_{Li} \psi_{Ci} S_{Qik}$$

(2) When the combination is controlled by the permonent load effect:

$$S_d = \sum_{j=1}^{m} \gamma_{Gj} S_{Gjk} + \sum_{i=2}^{n} \gamma_{Qi} \gamma_{Li} \psi_{Ci} S_{Qik}$$

where, γ_{Gj}—partial safety factor for the permanent load of number;

γ_{Q1}—partial safety factor for the variable load of number;

S_{Gjk}—load effects values are calculated in accordance with the characteristic values of permanent load G_{jk};

S_{Qik}—load effects values are calculated in accordance with the characteristic values of rariable load G_{ik};

ψ_{Ci}—coefficients of combination values of variable loads Q_i shall be adopted in accordance with the stipulations of the Clauses in Chapters of Code respectively;

m—number of the permanent loads participated in combinations;
n—number of the variable loads participated in combinations.

EXAMPLE 1.3

The self-weight of a working platform in a factory is 5.4kN/m², its live load is 2.0kN/m². The design service life is 50 years. Determine the design value S_d according to the fundamental combination.

SOLUTION

(1) If the situation controlled by the variable loads effects
$S_d = 1.2 \times 5.4 + 1.4 \times 1.0 \times 2 = 9.28 \text{kN/m}^2$
(2) If the situation is controlled by permanent load effect
$S_d = 1.35 \times 5.4 + 1.4 \times 1.0 \times 0.7 \times 2 = 9.25 \text{kN/m}^2$
Note: Though the dead load value is 2.7 times the value of live load, but from the two results, we can know the situation is controlled by live load.

EXAMPLE 1.4

A cantilever beam with T shape is subjected to liner distributed dead load g_k = 15kN/m; local live load q_k = 6kN/m, concentrated dead load at the end point: P_k = 20kN. Determine the maximum moment value M_A at the root section of the cantilever.

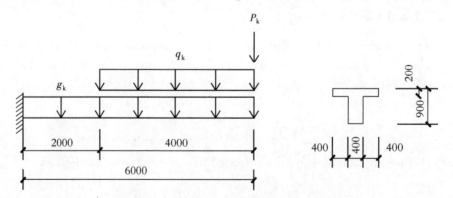

FIGURE 1.10　Wind load on shear wall structure

SOLUTION

(1) Calculation of the moment under characteristic value of different loads
Distribution load: $M_{qk} = g_k l^2/2 = 15 \times 6^2/2 = 270 \text{kN} \cdot \text{m}$
Concentrated load: $M_{qk} = P_{k1} = 20 \times 6 = 120 \text{kN} \cdot \text{m}$
Local live load: $M_{qk} = 6 \times 4 (6 - 4/2) = 96 \text{kN} \cdot \text{m}$

(2) Calculate fundamental combination

If the case is controlled by the variable loads effects:

$M_A = 1.2 \times (270 + 120) + 1.4 \times 1.0 \times 96 = 602.4 \text{kN} \cdot \text{m}$

If it is controlled by permanent load effect:

$M_A = 1.35 \times (270 + 120) + 1.4 \times 1.0 \times 0.7 \times 96 = 620.6 \text{kN} \cdot \text{m}$

So the unfavorable value is 620.6kN · m.

1.5 The Design Process

The design process is a sequential and iterative decision-making process. The three major phases are the following (Ref. 1.5, Ref. 1.6):

1.5.1 Definis Client's Needs and Priorities

All buildings or structures are built to fulfill a need. It is important that the owner or user is involved in determining the attributes of the proposed building. These include functional requirements, aesthetic requirements, and budgetary requirements. The latter include initial cost, premium for rapid construction to allow early occupancy, maintenancecost, and other life-cycle costs.

1.5.2 Develop Project Concept

Based on the client's needs and priorities, a number of possible layouts are developed. Preliminary cost estimates are made, and the final choice of the system to be used is based on how well the overall design satisfies the client's needs within the budget available. Generally, systems with simple concept and standardized geometries and details can take standard construction and cost lower investment. At this stage, the overall structural concept is selected. From approximate analysis of the moments, shears, and axial forces, preliminary member sizes are selected for each potential scheme. Once this is done, it is possible to estimate costs and select the most desirable structural system. The overall thrust in this stage of the structural design is to satisfy the design criteria dealing with appropriateness, economy, and, to some extent, maintainability.

1.5.3 Design Individual Systems

Once the overall layout and general structural concept have been selected, the structural system can be designed. Structural design involves three main steps. Based on the preliminary design selected in phase 2, a structural analysis is carried out to determine the moments, shears, torques, and axial forces in the structure. The individual members are then proportioned to resist these load effects. The proportioning, sometimes referred to as member design, must also consider overall aesthetics, the constructability of the design, coordination with mechanical and electrical systems, and the sustainability of the final structure. The final stage in the design process is to prepare construction drawings and specifications.

References

1.1 Technical specification for concrete structures of tall buildings[S]. JGJ 3—2010. Beijing: China Architecture and Building Press, 2011.

1.2 Nilson A H, Darwin D, Dolan C W. Design of Concrete Structures[M]. 14 Edition. NewYork: McGraw-Hill, 2009.

1.3 Load code for the design of building structures[S]. GB 50009—2012. Beijing: China Architecture and Building Press, 2012.

1.4 Code for design of concrete structures[S]. GB 50010—2010. Beijing: China Architecture and Building Press, 2010.

1.5 Unified Standard for reliability design of building structures[S]. GB 50068—2001. Beijing: China Architecture and Building Press, 2012.

1.6 Gu Xianglin. Building concrete structure design [M]. Shanghai: Tongji University Press, 2011.

1.7 Qiu Hongxing. Building structure design[M]. Beijing: Higher Education Press, 2013.

Questions

1.1 The self-weight of a working platform in a factory is $4kN/m^2$, its live load is $0.5kN/m^2$. The design service life is 50 years. Determine the design value S_d according to the fundamental combination.

1.2 Explain the categories and content of terrain roughness when calculating wind load according to Chinese load code.

1.3 For the ultimate limit states, if it is controlled by the variable loads effects, write down the equation for S (the design value of combination of load effects) and the meaning of each item.

1.4 What are Design reference period, Combination value and Characteristic value?

Chapter 2

Slabs

2.1 Introduction

Reinforced Concrete Slabs are very important structural components of buildings. It accounts for nearly 30% of the whole construction cost and controls about half of the total weight of buildings. Therefore, it is important to choose proper methods and design it properly.

Slab belongs to horizontal structure system, and together with vertical and lateral load resisting components forms an integral space structural system. It transfers the floor vertical load to vertical components (columns, walls etc.) and horizontal load to horizontal components (beam, girders etc.)

2.1.1 Types of Floor Systems

Slabs are classified into various categories.

2.1.1.1 Classfication according to the support conditions

According to support conditions, slabs can be classified as ribbed slab, flat slabs and waffle slab, with single span or several spans as shown in Figure 2.1. For ribbed slab, it can also divided into one-way or two-way span by the ratio of the side length in different directions or support condition.

For the ribbed slab, if it is supported by four edges of beams or walls, the vertical load on the slab are mainly transferred to the support by the bending effect. According to the theory of elastic thin plate, when the ratio of the longer side to the shorter side exceeds a certain value, the slab is one-way. In this case vertical loads are mainly transferred to the shorter side and loads transferred to the longer side are negligible. It is called a two-way slab. The force analysis can be explained below.

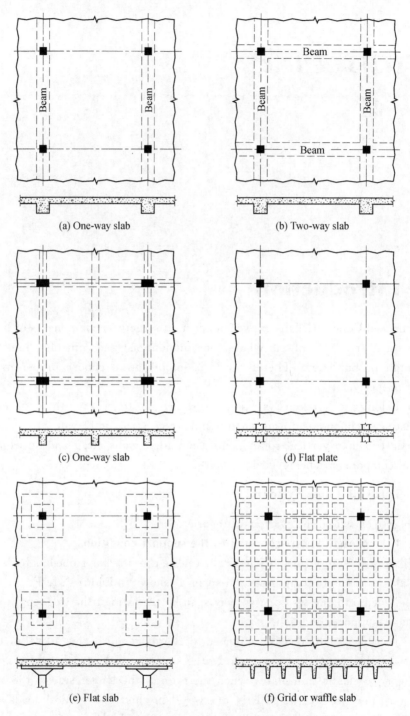

FIGURE 2.1 Normal types of floor systems

As shown in Figure 2.2, the rectangular slab is simply supported by the four edges and subjected to a vertical uniform load q. The slab's longer span is l_{01}, the shorter span is l_{02}. The two strips along x and y directions are taken out for calculation, which are suppose to be simply supported beam with on unit length, as shown in Figure 2.2. According to the deformation compatibility condition of the slab's mid-span:

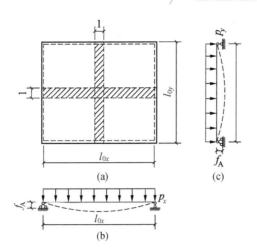

FIGURE 2.2 Loads' transmission on a quadrangular support slab

$$f_\Delta = \alpha_1 \frac{q_1 l_{01}^4}{EI_1} = \alpha_2 \frac{q_1 l_{02}^4}{EI_2} \tag{2-1}$$

where, α_1, α_2—deflection factor, when the ends are simply supported, $\alpha_1 = \alpha_2 = \frac{5}{384}$;

I_1, I_2—moment of inertia in x, y direction respectively;

l_{01}, l_{02}—calculated length in x, y direction respectively.

The total deflection at middle point is the sum of the deflections in x, y directions:

$$q = q_1 + q_2 \tag{2-2}$$

Assuming $I_1 = I_2$, according to formula (2-1) and (2-2), the load in x, y direction can be deduced by the deflection in x, y direction l_{01}, l_{02} as the following:

$$q_1 = \frac{l_{02}^4}{l_{01}^4 + l_{02}^4} q, \quad q_2 = \frac{l_{01}^4}{l_{01}^4 + l_{02}^4} q \tag{2-3}$$

if $l_{01}/l_{02} > 2$, it is called one-way slab. In this case, loads transferred to long edges are less than 5.9% of that of short edge. Otherwise, when the length ration of the longer edge to shorter edge is less than 2, it is called a two-way slab. In this case, the loads transferred to the shorter edges can not be neglected. According to Concrete Structure Design Code regulations:

(1) Slab supported by two opposite sides shall be calculated as one-way slab.

(2) Slab supported by four sides shall be calculated according to the following stipulations:

① If the ratio of the longer edge to shorter edge is greater or equal to 3.0, it should be calculated according to one-way slab.

② If the ratio of longer edge to shorter edge is less than or equal to 2.0, it shall be calculated according to two-way slab.

③ If the ratio of longer edge to shorter edge is greater than 2.0 and less than 3.0, it may be calculated as two-way slab. It can also be calculated according to one-way slab, but sufficient steel reinforcements shall be placed along longer edge.

Flat slab floor: Flat slab floor is a reinforced concrete slab which is supported by concrete columns without beams. They are often used in libraries and warehouses, etc.

Waffle slab floor: Waffle slab floor is divided into many small unites by dense ribs in two perpendicular directions. It has less weight compared with other slab types and usually used for large span floor such as the big halls of public buildings.

2.1.1.2 Classification according to the construction technique

According to different construction techniques, slabs can be divided into cast-in-place, precast slabs and fabricated monolithic slabs.

Cast-in-place slabs have good integrity and anti-seismic performance. The slab can be placed flexibly and has strong adaptability. However, the disadvantage of cast-in-situ is that the speed of construction is lower for much field work.

Precast slabs are constructed by prefabricated components, which are convenient for mechanized construction and thus shorten the construction period. However, the integrity and water resistance is poor and is inconvenient for setting openings. Precast floors are mostly used in simple and regular industrial buildings.

Fabricated monolithic slab: when the prefabricated components are installed, cast-in-place concrete are poured on the slab surface to strengthen the integrity of the slab. It has the advantages of the above two slabs, but its construction is more complicated.

2.1.2 Size of Reinforced Concrete Slabs

In the design of any reinforced concrete structural members such as floor slabs, beams and other structures, we can estimate in advance the sectional dimension based on experience and relevant data, given in Table 2.1. If the calculated results differ greatly to the estimated size, it should be reestimated until the requirements are met (Ref. 2.1).

TABLE 2.1 General dimensions of concrete beams and slabs

Categories		Height-span ratio (h/l)	Remarks
one-way slabs	simply supported continuous	$\geqslant 1/30$ $\geqslant 1/35$	The least thickness Roof slab: when $l<1.5$m, $h \geqslant 50$mm　　　when $l \geqslant 1.5$m, $h \geqslant 60$mm civil building floor: $h \geqslant 60$mm industrial building floor: $h \geqslant 70$mm floor under vehicle road: $h \geqslant 80$mm
two-way slabs	one-span and simply supported two-span and continuous	$\geqslant 1/40$ $\geqslant 1/45$(short span)	the slab's thickness: $80\text{mm} \leqslant h \leqslant 160\text{mm}$
waffle slabs	one-span and simply supported multi-span and continuous	$\geqslant 1/20$ $\geqslant 1/25$ (h is the height of rib)	the slab's thickness: $h \geqslant 50$mm the rib's height: $h \geqslant 250$mm
cantilever slabs (fixed ends)		$\geqslant 1/12$	when the cantilever length of the beam$\leqslant 500$mm, $h \geqslant 60$mm when the cantilever length of the slab>500mm, $h \geqslant 80$mm

Continued

Categories		Height-span ratio (h/l)	Remarks
flat slab	with column caps	$\geqslant 1/30$	$h \geqslant 150$mm
	without column caps	$\geqslant 1/35$	the width of the column cap $c=(0.2\sim 0.3)l$
Cast-in-place hollow floors		—	$h \geqslant 200$mm
multi-span continuous secondary beams		$1/18 \sim 1/12$	the beam's least height: secondary beams, $h \geqslant l/25$
multi-span continuous main beams		$1/14 \sim 1/8$	main beams, $h \geqslant l/15$
one-span simply-supported		$1/14 \sim 1/8$	the ratio of width to height is $1/3 \sim 1/2$, the modulus is 50mm

2.2 Design of One-Way Ribbed Slab

2.2.1 Structure Layouts of One-Way Slab

2.2.1.1 Compositions of floor systems

Floor systems consist of supporting members such as beams, columns and walls. Loads on building structures are transmitted to horizontal supporting members (beams) through slabs, and then transmitted to vertical members (columns or walls), and finally to the foundation.

2.2.1.2 Structural layout

One-way ribbed floor are composed of slab, the secondary beams and main beams. The secondary beams layout determines the slab panel dimensions, the main beams controls the span of the secondary beams and column networks which determine the span of the main beams. One-way slabs' spans are generally 1.7~2.5m, and should not exceed 3m. Secondary beams' spans are 4~6m and main beams' spans are 5~8m. (Ref. 2.2)

The layout of one-way slab should consider architectural effects, function and structural design principles. The common one-way slabs' layouts are the followings:

(1) The main beam is arranged transversely. The advantage is that the main beam and column can form a transverse frame with large lateral rigidity. Each frame is connected by a secondary beam, and the integrity of the building. (Figure 2.3a)

(2) The main beam is arranged in longitudinal direction. When the lateral spacing is larger than longitudinal spacing, the main beams can be arranged longitudinally. This can reduce the beam section height and increase the indoor height. (Figure 2.3b)

(3) Setting intermediate corridor without main beams. Suppose the corridor is in the middle of the plan, the main beams are supported by longitudinal internal load-bearing wall at the end. (Figure 2.3c)

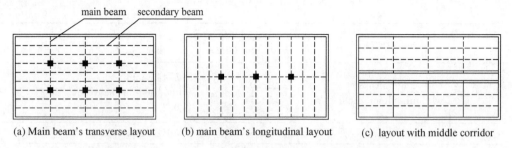

(a) Main beam's transverse layout (b) main beam's longitudinal layout (c) layout with middle corridor

FIGURE 2.3 Structure layout of one-way slab ribbed beam floor

2.2.2 Calculating Diagram of One-Way Slab

2.2.2.1 Structure simplification

For analyzing the structure, the following simplification can be carried out:

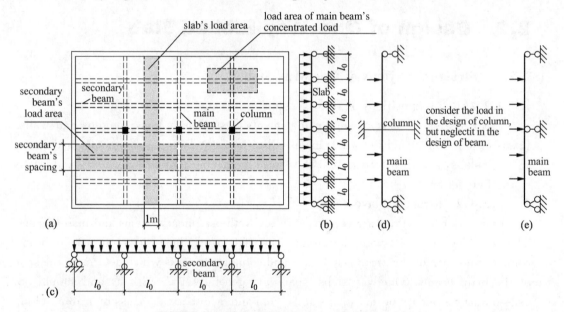

FIGURE 2.4 Calculating diagrams of one-way slab

(1) The calculation diagram of slab is supposed to be a 'continuous beam' which is simply supported by secondary beams. The calculation width of the slab is 1m and its linear load value is q on one unit area. (Figure 2.4b)

(2) The calculation diagram of secondary beam is supposed to be a 'continuous beam' which is simply supported by main beams. The load is transferred from the floor, and its self weight should be considered. (Figure 2.4c)

(3) When main beam and column are cast together into an integrity system, the internal force can be calculated according as a frame structure (Figure 2.4d); If beams flexural rigidity is much higher than columns (for example, the ratio of the stiffness between the main beam and the column is between $3 \sim 4$), the main beam can be regarded as a continuous beam hinged on the column (Figure 2.4e). Main beam resists dead loads and loads transferred from the secondary beam. Generally, loads of the main beam is much smaller than that transferred from secondary beam. In order to simplify

the calculation, it can be converted into concentrated load.

2.2.2.2 Calculation span

The calculation span of beam and slab is used for calculation of the internal forces. The effective span length l_0 is the distance between the rotation points of the two end supports. According to the elasticity theory, the span usually takes the distance between the two acting points of the support. According to the plasticity theory, the effective span is determined by the position of the plastic hinge. The specific values are shown in Table 2.2. l_0 is beam and slab's calculation span, l_n is beam and slab's clear span, a is beam and slab's supporting length, b is middle support's width.

TABLE 2.2 Calculation span of beam and slab

Calculation method	Beam and slab components		Effective span
Calculation by elastic theory	One span	Shelving at both sides	$l_0 = l_n + a$ $l_0 \leqslant l_n + h$ (slab) $l_0 \leqslant 1.05 l_n$ (beam)
		Shelving at one side and supporting component pouring at other side	$l_0 = l_n + a/2$ $l_0 \leqslant l_n + h/2$ (slab) $l_0 \leqslant 1.025 l_n$ (beam)
		supporting component pouring at both sides	$l_0 = l_n$
	Multi span	Side span	$l_0 = l_n + a/2 + b/2$ $l_0 \leqslant l_n + h/2 + b/2$ (slab) $l_0 \leqslant 1.025 l_n + b/2$ (beam)
		Middle span	$l_0 = l_c$ $l_0 \leqslant 1.1 l_n$ (slab) $l_0 \leqslant 1.05 l_n$ (beam)
	Support at both sides		$l_0 = l_n + a$ $l_0 \leqslant l_n + h$ (slab) $l_0 \leqslant 1.05 l_n$ (beam)
	Support at one side and supporting component pouring at other side		$l_0 = l_n + a/2$ $l_0 \leqslant l_n + h/2$ (slab) $l_0 \leqslant 1.025 l_n$ (beam)
	supporting component pouring at both sides		$l_0 = l_n$

2.2.3 Internal Forces Analysis by Elastic Theory Method

2.2.3.1 Most unfavorable positions of live loads

According to elastic theory method, the beams and slabs are supposed to be a ideal elastic system. By adopting the above calculation diagram, the internal force can be calculated out based on structural mechanics method.

The appendix B1 shows the internal force coefficient of 2~5 span of the continuous beam, when the beam or slab's span number is more than 5, it can be approximately taken as 5 spans; the internal forces are the same for the third span and other desirable middle spans in the calculation of the reinforcement. If spans of continuous beam are different, but less than 10%, it can be seen as equal spans. When calculating the negative moment of the support, the calculation span can take the average value of the adjacent two spans or the greater values, when calculating the mid-span moment the calculation span takes the length of corresponding span.

(1) In order to obtain the maximum positive bending moment of a span, the variable load should be arranged on the span, and then apply loads on every two span of both sides.

(2) In order to obtain the minimum bending moment of a span, the variable loads should be applied on the two adjacent spans and then on every two span of both sides.

(3) In order to obtain the maximum negative bending moment of a section of bearing, variable loads should be placed on two adjacent spans of the bearing, and then on every two span of both sides.

(4) In order to obtain the maximum shear force of a section of bearing, the layout of the variable load is the same as the maximum negative bending moment of the section.

According to the above principle, the most unfavorable arrangements of variable loads can be determined. Combined with the permanent load (all over the span) they make the most unfavorable load combination.

2.2.3.2 Envelope diagrams of internal forces

For a single span beam or slab, it is obvious that the maximum internal force will be generated when all permanent loads and variable loads are applied simultaneously. But for a specified section of a multi-span continuous beam or slab, the internal force caused by all the loads at the same time is not necessarily the largest. To make the continuous beam or slab can be designed reliably under various possible load arrangements, it is necessary to find the most unfavorable internal force that may be generated on each section. The most unfavorable position of the variable load must be considered.

Figure 2.5 shows the bending moment diagram and shear force diagram of a 5-span continuous beam of different spans with different loads. If the variable load acts at a certain span, its cross span bears positive bending moment, and the middle bending moment in the adjacent span bears negative bending moment. Therefore, it is not difficult to deduce the most unfavorable position of variable loads acting on a continuous beam or slab.

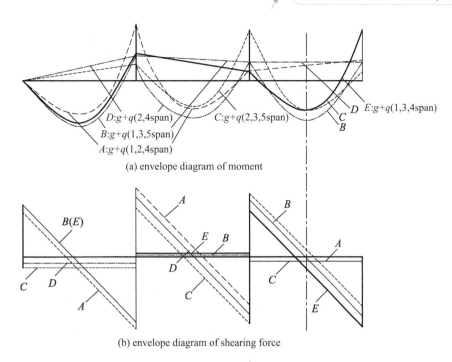

FIGURE 2.5 Continuous beam's internal force diagram

As shown in Figure 2.5, the outer lines form an envelope graph, which reflects the maximum internal force value that each section may produce, and is the basis of selecting section and layout of rebars in design.

2.2.3.3 Converted loads

In the calculation model of one-way slab and secondary beam, the supports are supposed to be ideal hinges. In fact, slab, secondary beam and main beam are cast into integrity. Taking the slab as an example, the rotation of slab will induce the rotation of secondary beam; in turn, the torsional resistance of secondary beam will limit the rotation of slab and reduce the angle of rotation ($\theta' < \theta$), θ is the angle of rotation of the hinge support as shown in Figure 2.6, this reduces slab internal force. In order to make the calculation of slab internal force close to the actual result, the converted load is used. The method is also applicable to secondary beams supported by the main beam. Since the rotation of slab and secondary beam bearings is mainly due to the unfavorable arrangement of live loads, the method keeps total load constant, but increases constant

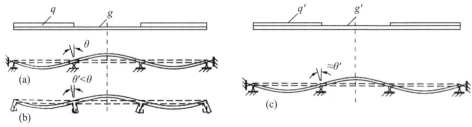

(a) Deformation of ideal hinge support   (b) Deformation of support under elastic constraints   (c) Deformation under converted load

FIGURE 2.6 The converted load of continuous beams and unidirectional continuous slabs

load and reduces live load.

Continuous beam $\qquad g'=g+\dfrac{q}{2}, \ q'=\dfrac{q}{2}$ (2-4)

Continuous slab $\qquad g'=g+\dfrac{q}{4}, \ q'=\dfrac{3q}{4}$ (2-5)

where, g, q—actual design value of dead load and live load on unit length;

g', q'—converted design value of dead load and live load on unit length.

When slab and secondary beam is placed on the masonry or steel structure and the supports can not confine their rotation, then load will not be converted. When the main beam is calculated as continuous beam, columns rigidity is small, then load shall not be converted too.

2.2.4 Internal Forces Analysis by Plastic Theory Method

2.2.4.1 Plastic hinges of concrete flexural member

As shown in Figure 2.7, a simply supported beam is subjected to a concentrated load. When the tensile rebar of the middle span yield with the increased load, the yield moment of the section is M_y, and the corresponding sectional curvature is φ_y. With the increasing load, the stress of the rebar remains the same, but the neutral axis moves upwards. When the force is increased slightly, the compression zone of concrete edge reaches the ultimate compressive strain, and bending moment increases to ultimate moment M_u, with the corresponding section curvature φ_u. It can be seen from the diagram, the increment of the moment $(M_u - M_y)$ is very small and the increase of sectional curvature $(\varphi_u - \varphi_y)$ is very large during the process. When moment basically remains unchanged, sectional curvature increases sharply, the section yields a certain degree of plastic rotation in the direction of the bending moment. Some parts of the adjacent cross section also yield along the direction of plastic moment rotation, as if there is a 'hinge' bearing moment, which is called plastic hinge in Engineering. (Ref. 2.3)

In Figure 2.7, the part that $M \geqslant M_y$ is the plastic hinge region. Which is the length of plastic hinge l_p, the angle θ_p is the rotation angle of the plastic hinge. In general, the region with intensive the plastic deformation is idealized as a plastic hinge which concentrates at the same cross section. The difference between plastic hinge and ideal hinge: ideal hinge cannot bear the bending moment, it is a two-way hinge and concentrates at one point; plastic hinge can bear the bending moment, it is a one-way hinge and distributes in a certain length.

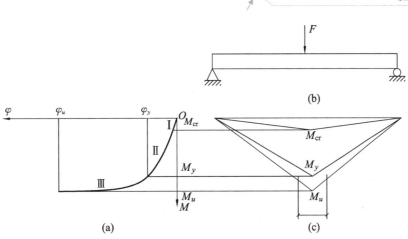

FIGURE 2.7 Bending moment diagram of simply supported beam under concentrated load

2.2.4.2 Internal force redistribution of statically indeterminate structures

For the statically indeterminate reinforced concrete structures, due to the change of stiffness and the occurence of plastic hinge, internal force are redistributed during the different sections. The results of internal force and deformation are different from that of elastic analysis. Here is an example.

EXAMPLE 2.1

There is two-span reinforced concrete continuous beam, the calculated spans are both l_0, each span bears a concentrated load P. Try to analyze the moment changed with the load at the section of intermediate support and where the load is applied.

SOLUTION

(1) Before plastic hinges are formed:

The moment diagram of the two-span continuous beam is shown in Figure 2.8a which is obtained by structural mechanics method, the maximum value of the moment is at the cross section of the middle support B. It is assumed that the intermediate cross section reaches its ultimate flexural capacity M_{By} under load P_1 when the plastic hinge is formed. P_1 can be determined by Equation (2-6):

$$M_{By} = \frac{3}{16} P_1 l_0 \tag{2-6}$$

At the same time, the bending moment of the cross section of point 1 is

$$M_{11} = \frac{5}{32} P_1 l_0 \tag{2-7}$$

At this moment, the beam does not lose its bearing capacity. Suppose the flexural capacity of section 1 is M_y, the flexural capacity of the section has $(M_y - M_{11})$ margin.

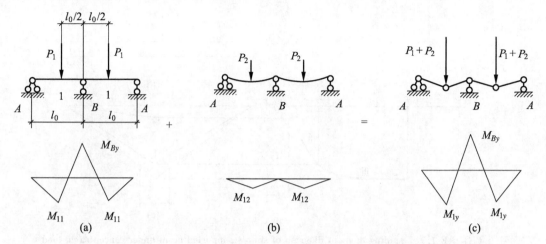

FIGURE 2.8 The bending moment of two span continuous beams changed with load

(2) After plastic hinges are formed:

After the plastic hinge is formed at the cross-section B of the intermediate support, the two-span continuous beam turns into two simple beams (Figure 2.8b). If the load increases, the incremental bending moment caused by incremental load P_2:

$$M_{12} = \frac{1}{4} P_2 l_0 \tag{2-8}$$

When the load point 1 also reaches its bending capacity M_{1y}, the whole structure changes to a geometrically unstable system and is destroyed. Assuming $M_{11} + M_{12} = M_{1y}$, the corresponding P_2 can be obtained. Therefore, the maximum value of the concentrated load that the continuous beam can bear is $P_1 + P_2$.

2.2.4.3 Method of moment modulation

The general method for calculating continuous beams and slabs' redistribution of internal force is the moment modulation method. The method of redistribution of plastic internal forces should not be taken into account for members subjected to direct dynamic loads or requiring no cracks.

The method of moment modulation adjusts the bending moment and shear force obtained by elastic method considering the influence of redistribution of internal force.

$$M = (1 - \beta) M_e \tag{2-9}$$

where, M—design value of bending moment after modulation;

M_e—design value of bending moment calculated by elastic method;

β—moment amplitude modulation coefficient of section.

For equal span continuous beam or continuous one-way slab under uniform load, the method can also be adopted. The internal force coefficients deduced by the method of amplitude modulation are listed into form and used to calculate the internal force for design. The bending moment at mid-span and support, and the design shear force at supports can be calculated according to Formula 2-10 and 2-11:

$$M = \alpha_M (g + q) l_0^2 \tag{2-10}$$

$$V = \alpha_V (g + q) l_n^2 \tag{2-11}$$

where, α_M—the moment coefficient considering redistribution of plastic internal forces, as shown in Table 2.3;

α_V—the shear coefficient considering the redistribution of internal forces, as shown in Table 2.4;

g, q—design values for uniformly distributed permanent and variable loads;

l_0—the calculation span, calculated by plastic theory method, as shown in Table 2.2;

l_n—clear span.

TABLE 2.3 Bending moment's calculation coefficient α_M of continuous beam and one-way slab

Support conditions		Sectional position				
		Side span	Middle side span	Second support apart from side span	Middle span	The middle of middle span
beam and slab are shelved on the wall		0	$\frac{1}{11}$	Two continuous span: $-\frac{1}{10}$ More than three continuous span: $-\frac{1}{11}$	$-\frac{1}{14}$	$\frac{1}{16}$
Slab	Connect with beam	$-\frac{1}{16}$	$\frac{1}{14}$			
Beam	cast-in-place	$-\frac{1}{24}$				
Beam connect cast-in-place with column		$-\frac{1}{16}$	$\frac{1}{14}$			

TABLE 2.4 Shear force's calculation coefficient α_V of continuous beam

Support conditions	Sectional position				
	Outside of side span	Second support apart from side span		Middle span	
		Outside	Inside	Outside	Inside
On the wall	0.45	0.60	0.55	0.55	0.55
Connect cast-in-place with beam and column	0.50	0.55			

The equal span continuous beam under concentrated load can be calculated by the following method. The bending moment at mid-span and the designing shear force at support's edge can be calculated according to the following formula:

$$M = \eta \alpha_M (G+Q) l_0 \qquad (2\text{-}12)$$
$$V = \alpha_V n (g+q) l_n \qquad (2\text{-}13)$$

where, α_M—the moment coefficient considering redistribution of internal forces, as shown in Table 2.3;

α_V—the shear coefficient considering the redistribution of internal forces;

η—correction factor of concentrated load;

G, Q—design values for concentrated permanent and variable loads;

n—the number of span's concentrated loads.

2.2.5 Section Design and Reinforcements Construction of One-Way Slabs

After obtaining beam and slab's internal force, the bearing capacity and the reinforcement calculation of the section can be carried out according to the internal force. If the component satisfies the construction requirements of Table 2.1, deformations and cracks are usually satisfied without calculation.

2.5.5.1 Design of slab section

(1) Bending moment design value of continuous slab

The continuous beam is integrally connected with slab. The middle-span section will crack at the bottom under the positive sagging moment and the support section will crack in the upper part under the negative moment, thus the stresses in the slab forms an arch (Figure 2.9). The beam in horizontal direction applies constraint on slab, produces arch effect and reduces the moment value of slab.

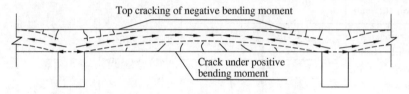

FIGURE 2.9　Arch action of one-way continuous slab

Considering the beneficial effect, due to related codes, no matter by elastic or plastic theory, the moment value of the slab grid that connects with beams along the perimeter shall be deduced as the following:

① decrease 20% for mid−span section and mid-support section.

② The percent of side-section's mid-span section and the second span section from floor edge: when $l_b/l < 1.5$, it is 20%; when $l_b/l \leqslant 2$, it is 10%. l is the calculating span perpendicular to floor's edge, l_b is the calculating span along floor's edge (Figure 2.10)

③ no reduction for floor angle grid.

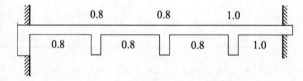

FIGURE 2.10　Bending moment reduction coefficient of one-way continuous slab

(2) Reinforcement calculation

The section of the slab is calculated, and the strip with 1m width is designed as a singly reinforced rectangular section. The reinforcement of the short span of one-way slab is determined by calculation, and the reinforcement along the long span can be calculated according to the construction requirement. Generally, the slab can satisfy the requirement of shear carrying capacity of the inclined section, and the shear capacity does not need calculation for design.

2.2.5.2 Construction requirements of slabs

(1) Slab's thickness

Slab's thickness h shall meet requirements of Table 2.1. In order to deduce costs and structure's weight, slab should not be too thick.

(2) Slab's reinforcement construction

Stress steel bar usually adopts HPB300 grade with diameter 6mm, 8mm, 10mm or 12mm; the negative bending moment of the bearing should not be too small. Selected steel bars should not be more than two diameters. The spacing of the steel bars in the slab shall not be less than 70mm; when $h \leqslant 150$mm, the space should be $\leqslant 200$mm; when $h > 150$mm, it should be $\leqslant 1.5h$ and $\leqslant 250$mm.

In addition to the steel rebars in short span direction, distributed vebars should also be placed in the direction perpendicular to the main rebars. The sectional area of the distribution reinforcement shall not be less than 15% of the calculated steel bar, and the reinforcement ratio shall not be less than 0.15%. The distribution reinforcement diameter should not be less than 6mm, the spacing should not be greater than 250mm. When the concentrated load is large, the area of the distribution reinforcement should be increased, and the spacing should not be greater than 200mm.

When the slabs are parallel to the reinforcing bar along the beam direction, it should be equipped with additional reinforcement, the number should not be less than 1/3 of the slab steel reinforcement, the diameter should not be less than 8mm, and the spacing should be less than 200mm. From of the edge of the main girder, the length of the rebar should not be less than $l_0/4$. (Ref. 2.1)

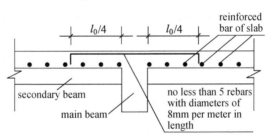

FIGURE 2.11 Arangement of rebars in slab along main beam

For cast-in-place concrete slab simply supported or embedded in the load-bearing masonry walls, top constructional reinforcement should be placed along supporting perimeter, the diameter should be $\geqslant 8$mm, and the space should $\leqslant 200$mm. the length of the distributed vebars should not be less than 1/7 that of the short span l_y (Figure 2.12). For the corner area, the embedded reinforcement shall be arranged in the upper part of the slab along both sides, and the length that extends wall shall not be less than $l_y/4$ (Figure. 2.12).

For the cast-in-place slab, the sectional area of the upper structural steel bars (including bent reinforcing bars) embedded in the slab arranged in the direction of short span should not be less than 1/3 of the reinforced steel bar of the mid-span.

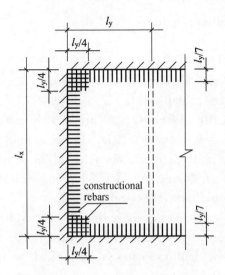

FIGURE 2.12 Constructional reinforcement embedded at the top of slab

(3) Reinforcement layout mode

The arrangement of reinforcement in the continuous slab has two modes: bent or separated (Figure 2.13). The bent reinforcing bars (Figure 2.13a, b) that bearing positive moment are bent at proper position and extend across the supports to bear negative

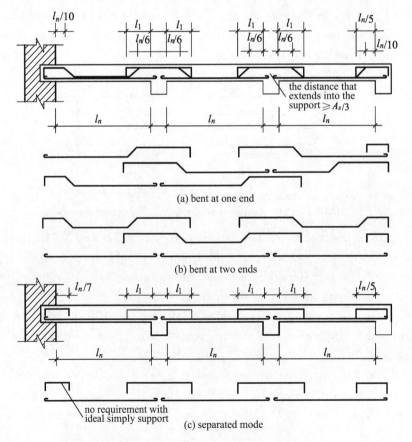

(a) bent at one end

(b) bent at two ends

(c) separated mode

FIGURE 2.13 Reinforcement of continuous slab

moment. In addition, the distance of other rebars without bending that bearing positive moment shall not be greater than 400mm, and the sectional area shall not be less than 1/3 of the cross section of the middle positive moment steel bars in this direction. The bending angle of bent steel bars is generally 30°. When the thickness of the slab $h >$ 120mm, it is 45°. Generally, 1/3 to 2/3 rebars that bearing the mid-span positive moment should be bent from the point that is $l_n/6$ from the support edge to resist the negative moment. If the sectional area of the steel bar does not meet the requirements of the support section, additional reinforcement can be added. The separated reinforcing bars (Figure 2.13c) are that all the mid-span positive moment steel bars extend into the support, and the negative bending moment steel bars are arranged individually.

Compared with the separated rebar, the bent reinforcing bar be better anchored and saves steel, but the construction is more complicated. The anchor performance of separate steel bar is bad, but its design and construction is convenient, and normally used in real projects. When the thickness of the slab exceeds 120mm and the dynamic load is large, the bent reinforcement can be used.

2.2.5.3 Design of beam sections

For the beam in the floor of one-way ribbed slab, the bearing capacity and reinforcement of the normal section and the inclined section should be calculated according to the internal forces. The mid-span section bearing the positive bending moment should be calculated as T-Section. The flange width b'_f value is calculated according to the provisions of the GB 50010 specification. The bearing section subjected to negative bending moment is calculated as rectangular section with width b.

Beams generally adopt HRB335 and HRB400 grades to resist forces in the longitudinal direction. When beam's depth $h=300$mm, the diameter of steel should not be less than 10mm, when the beam's depth is greater than 300mm, it should not be less than 8mm.

At the support of the main beam, due to the negative moment steel bars of slab, secondary beam and main beam overlap each other, main beam steel bars are generally put under secondary beam reinforcement, thus reduce the effective height of main beam cross-section. In the calculation of bearing capacity of the main beam section at the support, the effective height of the cross section h_0 is generally taken as (Ref. 2.4):

Single row steel bar: $h_0 = h - (50 \sim 60 \text{mm})$ (2-14)

Double row steel bar: $h_0 = h - (70 \sim 80 \text{mm})$ (2-15)

where, h—the beam cross section height.

2.2.5.4 Constructional requirements of beams

The section size of beam can be estimated by reference to Table 2.1. For longitudinal rebar, the minimum reinforcement ratio, the anchorage length, bending position, the position of the rebar that can be cut off, stirrup diameter, spacing and minimum stirrup ratio requirements can be found in the relevant provisions of the GB50010 specification, see Appendix B.

Truncation and bending of rebars of secondary beams, in principle, should be determined according to the internal force envelope diagram. For secondary beam, if it is equal span or the difference of span is less than 20%, when the variable load and permanent load ratio $q/g \leqslant 3$, the reinforced truncation and bending position are made according to the structure regulations shown in Figure 2.14. l_{ab} is the basic anchorage length of the stressed steel bar.

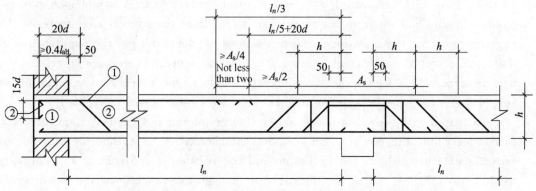

(Rebars only need to extend to the beam end if it can move freely embedded in wall.)

FIGURE 2.14 Constructional reinforcement regulation of secondary beam

At the intersection of main beam with the secondary beam, the main beam suffers concentrated load transferred by secondary beam. If the secondary beam is located at the lower part of beam section height or within the scope of the concentrated load, it should be reinforced with additional transverse reinforcement (stirrup, suspension bar). The stirrups should be arranged in the range of $l_s = 2h_1 + 3b$ (Figure 2.15).

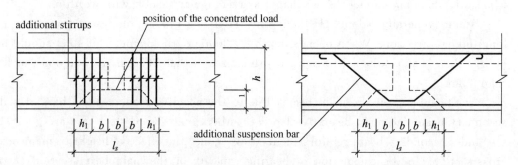

FIGURE 2.15 Layout of additional transverse reinforcement under concentrated load

The total cross-sectional area of additional transverse reinforcement can be calculated by:

$$A_{SV} = \frac{P}{f_{yv} \sin\theta} \tag{2-16}$$

where, A_{SV}—total cross-sectional area of additional transverse reinforcement required for concentrated load; When using additional suspension bars, A_{SV} is the total section area of left and right bending rebars.

P—design value of concentrated load of main beam transferred by secondary beams.

f_{yv}—design value of tensile strength of steel bars.

θ—the angle between the additional transverse reinforcement and the beam axis.

2.3 Design of Two-Way Ribbed Slab

2.3.1 Structure Layouts of Two-Way Ribbed Slab

The structural layout of two-way ribbed slab is similar to that of one-way slab. The difference is that primary and secondary beams are not obvious, and no longer described here.

2.3.2 Stress Characteristics of Two-way Ribbed Slab

In the section 2.1 of this chapter, the one-way slabs and two-way slabs are distinguished from the force analysis.

Figure 2.16 shows the failure characteristics of a simply supported two-way slab. When the load increases gradually, cracks firstly occur at the middle of the slab, the first crack appears at the bottom along the long side direction; when the load continues to increase, these cracks extend gradually and travel along 45-degree line towards the four corners; just before failure, the circular cracks appear at the top face of the slab near the four angles, cracks continue expanse along the diagonal direction of slab's bottom, and then finally collapse due to the yield of steel bars at four edges of the middle of the slab.

Under the action of the load, the four edges of slab may crul up, the pressure of the four supporting beam transferred form the slab in not evenly distributed. The value is large at the the middle and small at the ends. (Ref. 2.5). If the angle of slab is constrained by the wall, it is more likely that circular or linear oblique cracks will appear at the top of slab, as shown in Figure 2.16. How to avoid oblique cracks is one of the important contents in the design of slab structures in practical engineering.

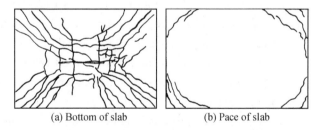

(a) Bottom of slab   (b) Pace of slab

FIGURE 2.16 Failure characteristics of simply supported two-way slab

2.3.3 Analysis of Two-Way Slab's Internal Forces by Elastic Theory Method

2.3.3.1 Internal forces calculation of one unit two-way slab

The calculation method of two-way slab based on elastic theory belongs to the bending problem of thin slab with small defection. As the internal force analysis is very complicated, in practical design, in order to simplify the calculation, internal forces are calculated directly based on reference forms deduced from the elastic theoretical analysis results. In the appendix table of the book, six calculation diagrams are selected with different boundary conditions. The internal bending moment coefficients (Poisson's ratio $v=0$), the moment coefficient of the support and the deflection coefficient are given respectively under uniformly distributed loads. The moment and deflection can be calculated by Equation 2-17 and 2-18.

$$M = \alpha_M \times (g+q)l^2 \qquad (2\text{-}17)$$

$$f = \alpha_f \times \frac{(g+q)l^4}{B_c} \qquad (2\text{-}18)$$

where, M— maximum bending moment design value of one-unit width central slab strip with inner span or at the support.

f—maximum deflection of the center slab strip.

α_M—bending moment coefficient.

α_f—deflection coefficient.

g, q—design value of uniform load and live load on slab.

B_c—bending rigidity of the slab strip section.

l——the smaller value between l_{ox}, l_{oy} which are span in two directions.

The effect of transverse deformation should be taken into account for the internal bending moment.

$$m_x^v = m_x + v m_y \qquad (2\text{-}19)$$

$$m_y^v = m_y + v m_x \qquad (2\text{-}20)$$

where, m_x^v, m_y^v—considering the influence of v, the mid-span bending moment in two directions.

m_x, m_y—when $v=0$, the mid-span bending moment in two directions.

v—Poisson's ratio, for reinforced concrete, $v=0.2$.

2.3.3.2 Internal force calculation of Equal-span Multi-units Continuous two-way slabs

It is more complex to calculate the internal force of continuous two-way slabs. The following calculation method is usually used when it is equal-span or span difference is less than 20% in the same direction. This method assumes that the flexural stiffness of slab and the deformation of the beam are negligible; torsional stiffness of the supporting beams is very small and can be rotated. The most unfavorable layout of the live load on the two-way slab and the supporting condition are simplified reasonably, and the multi-units continuous slab is taken as a one unit slab for calculation.

(1) Calculation of maximum bending moment of the middle span of the slab units.

When calculate the maximum bending moment of mid span of one slab unit, live load should be applied on this unit, and also on every two spans alternately, which is also called checkerboard layout. (Figure 2.17a). In order to make use of single-span two-way slab internal force calculation coefficient table, the variable load can be decomposed into symmetric load $q/2$ (Figure 2.17b) and antisymmetric load $q/2$ (Figure 2.17c).

For multi-uint bidirectional continuous slab, under the load effect of symmetrical load $g' = g + q/2$, the load on both sides of all intermediate supports are the same, the rotation angle at the section of supports can be taken as zero, and the middle unit are supposed to be two-way slab fixed by four edges.

Under antisymmetric loads $q' = \pm q/2$, the rotation angles of adjacent slab units have same value and direction, and the middle units can be taken as simply supported slab by the four edges.

For the slab units close to the edges or at the corner, the maximum moment at the corner can also be calculated by the above method, but the outer boundary conditions should be determined according to the actual situation.

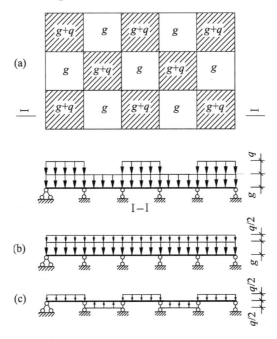

FIGURE 2.17 Failure characteristics of two-way slab simply supported by four edges

(2) Calculation of supports' maximum bending moment

In order to simplify the calculation, it is assumed that the bending moment has the maximun value when the permanent load and the variable load distribute uniforming on the continuous two-way slabs, and the intermediate supports are fixed. As for the side and corner zones, the outer boundary condition should be considered according to the actual situation.

For the intermediate support, the moment values of the supports of the adjacent two zones are often not equal, and the average value can be approximately taken into account in the calculation of the reinforcement.

2.3.4 Internal Forces Analysis of Two-Way Slab using the Plastic Theory Method

According to the plastic theory, there are two common methods of calculating the internal force of two-way slabs: The plastic hinge line method and slab strip method. This book only introduces the method of plastic hinge line.

2.3.4.1 The concept and location of yield line

For a simply supported beam, when the tensile steel yields at the middle section of the beam, plastic hinge produces in the beam. However, for a simply supported two-way reinforced concrete slabs, the plastic hinges will appear at the mid-span section when the tensile steel bars yield. The bending moment on the plastic hinge line is considered as the ultimate bending moment, and the corresponding load is the ultimate load. With the increase of load, the plastic line will continue to appear at the crak of the upper and lower surfaces in the slab, the plastic internal force redistribute in the cross section until the slab forms the failure mechanism, at this time, the plastic hinge line will be divided into some blocks. The plastic hinge line method of solving the equation is based on the failure mechanism of the slab, the ultimate bending moment and reinforcement can be calculated out by the method. Specifically, the plastic hinge method of calculating the two-way slab step is divided into three steps:

(1) Suppose that various failure mechanisms of a slab are geometrically variable systems, which are divided by a series of plastic hinge lines.

(2) Establish a virtual displacement mechanism, calculate the corresponding angles of each plastic hinge line and the virtual displacement of the slab according to the geometric conditions.

(3) Based the principle of virtual work, establish the relationship between external load and bending moment on plastic hinge line.

The position of the plastic hinge line is related to the plane shape of the slab, the boundary conditions, the load form, the condition of the reinforcement, and so on. The plastic hinge line at the top of the plate bearing positive bending moment is called the negative plastic hinge line. The plastic hinge line at the bottom of the plate bearing positive bending moment is called the positive plastic hinge line. Usually, the negative plastic hinge line occurs at the fixed boundary and the positive plastic hinge line passes through the intersection of the rotating axes of the adjacent blocks and has the maximum bending moment. Figure 2.18 shows the plastic hinge distribution mode of some common slabs.

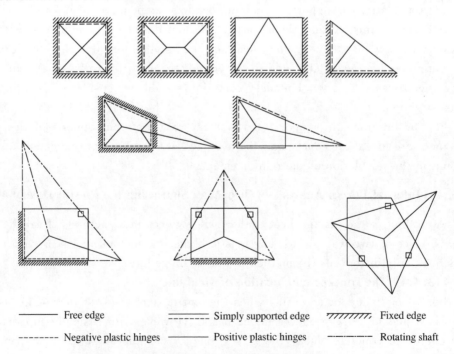

FIGURE 2.18 Plastic hinge line of slab

2.3.4.2 Basic assumptions of yield-line theory

When the two-way ribbed slab is calculated by the plastic hinge line method, it follows assumptions below:

(1) Before the slab breaks, the plastic hinge line occurs at the maximum bending moment.

(2) Form a flexible variable system with plastic hinge line (failure mechanism).

(3) Under distributed load, the plastic hinge line is straight.

(4) The plastic hinge line divides the slab into several parts, which can be regarded as rigid, and the deformation of the whole slab is concentrated on the plastic hinge line. The slabs rotate around the plastic hinge line when they are destroyed.

(5) There is only a certain value of ultimate bending moment on the plastic hinge line, where the value of torque and shear force can be considered to be zero.

2.3.4.3 The bending moment modulation method of continuous beam and continuous unidirectional slab

At present, bending moment modulation method is commonly used method for redistribution of plastic internal forces in engineering. The bending moments are first calculated by elasticity method, then select some moment values that first appear at plastic hinge section, adjust the values according to the principle of internal force redistribution, and then perform reinforcement calculation. The amplitude of section bending moment can be adjusted by the amplitude modulation factor β_M

$$M = (1 - \beta_M) M_e$$

where, M—adjusted bending moment design value;

M_e—the design value of the bending moment value calculated by the elastic method.

2.3.4.4 The determination of failure mechanism

In order to determine the failure mechanism of plates, it is necessary to find the location of plastic hinges. The location of the plastic hinge line can be determined according to the following four principles:

(1) The symmetrical structure has a symmetrical plastic hinge line distribution.

(2) The positive plastic hinge appears at the positive bending moment area and the negative plastic hinge appears at the negative bending moment area.

(3) The number of plastic hinge lines should make the whole slab become a geometrically variable system.

(4) The plastic hinge line should meet the requirements of rotation.

Each plastic hinge is a boundary between two adjacent rigid slabs and should rotate with two adjacent blocks so that the plastic hinge line passes through the intersection of the adjacent slabs.

When studying the multi-failure mechanism, it is necessary to obtain the minimum limit load. When different failure mechanisms are used with multiple variables, the ultimate load can be obtained by the ultimate load of the variable differential method.

2.3.4.5 Calculation of internal force of single two-way slab under uniform load

As shown Figure 2.19 a rectangular fixed two-way slab is subjected to the uniform load p, its short side and a long side are l_{01} and l_{02} respectively. From engineering practice, the failure pattern can be approximately assumed as shown in Figure 2.19. Negative plastic hinge lines formes at four supporting edges. The positive plastic hinge line forms a symmetrical pattern and develops along the 45-degree direction. This greatly simplifies the calculation and the error is within 5%. At this point, the plastic hinge line will divide the slab into four pieces. According to the principle of virtual work, external work done by the external force should be equal to the work done by the internal force. The internal and external force balance conditions are also considered.

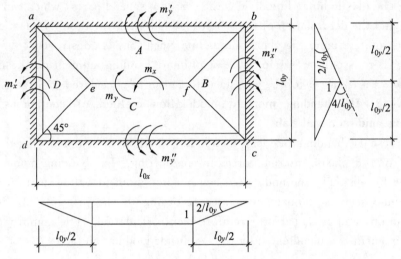

The failure mechanism of two-way slab which is fixed by four sides under uniform load

FIGURE 2.19 Four-sided fixed or continuous two-way plastic hinge line and free body diagram

If the vertical displacement of the failure mechanism is unit 1, the work under the uniform load $(g+q)$ is:

$$W_{ex} = (g+q)\left[\frac{1}{2} \times l_{0y} \times 1 \times (l_{0x} - l_{0y}) + 2 \times \frac{1}{3} \times l_{0y} \times \frac{l_{0y}}{2} \times 1\right]$$

$$= (g+q)\frac{l_{0y}}{6}(3l_{0x} - l_{0y}) \tag{2-21}$$

According to Figure 2.19, the rotation angle of negative plastic hinges is $\frac{2}{l_{0y}}$. On the normal plastic hinge line ef, the relative rotation angle of plate A and C is $\frac{4}{l_{0y}}$. The rotation angles of oblique normal plastic hinge along long span and short span directions are both $\frac{2}{l_{0y}}$. Therefore, the internal work of the ultimate bending moment on the negative plastic hinge line is:

$$[(m'_x + m''_x)l_{0y} + (m'_x + m''_x)l_{0y}]\frac{2}{l_{0y}}$$

The internal work of the ultimate bending moment on the normal plastic hinge line ef is:

$$m_y(l_{0x} - l_{0y})\frac{4}{l_{0y}}$$

Internal work of the ultimate bending moment along the long span direction of the four oblique normal plastic hinges is shown as following:

$$4m_x \frac{l_{0y}}{2} \cdot \frac{2}{l_{0y}} = 4m_x$$

Internal work of the ultimate bending moment along the short span direction of four oblique normal plastic hinges is shown as following:

$$4m_y \frac{l_{0y}}{2} \cdot \frac{2}{l_{0y}} = 4m_y$$

The total internal work of the ultimate bending moment on the plastic hinge line is shown as following:

$$W_{in} = [(m'_x + m''_x) l_{0x} + (m'_y + m''_y) l_{0x}] \frac{2}{l_{0y}} + m_y (l_{0x} - l_{0y}) \frac{4}{l_{0y}} + 4(m_x + m_y)$$

$$= [2(m_x l_{0y} + m_y l_{0x}) + (m'_x l_{0y} + m''_x l_{0y}) + (m'_y l_{0x} + m''_y l_{0x})] \frac{2}{l_{0y}} \quad (2\text{-}22)$$

According to the principle of virtual work, when a failure mechanism is formed, the external work done by the ultimate uniform load $(g+q)$ should be equal to the internal work done by the ultimate bending moment on the plastic hinge. The assumption is shown as following:

$$M_x = m_x l_{0y}, M'_x = m'_x l_{0y}, M''_x = m''_x l_{0y}$$
$$M_y = m_y l_{0x}, M'_y = m'_y l_{0x}, M''_y = m''_y l_{0x}$$

The basic formula for two-way slab by plastic hinge method is shown as following:

$$2M_x + 2M_y + M'_x + M''_x + M'_y + M''_y = \frac{1}{12}(g+q) l_{0y}^2 (3l_{0x} - l_{0y}) \quad (2\text{-}23)$$

Suppose:

$$m_x / m_y = \alpha_m, n_l = l_{0x}/l_{0y}, m'_x / m_x = m''_x / m_x = m'_y / m_y = m''_y / m_y = \beta_m$$

We can get:

$$m'_x / m''_y = \beta_m m_x, m'_y / m''_y = \frac{\beta_m}{\alpha_m} m_x, m_x = m_x l_{0x}/n_l, m_y = m_x l_{0x}/\alpha_m$$

Substitute them into the basic formula for two-way slab:

$$\frac{1}{12}(g+q) l_{0y}^2 (3l_{0x} = l_{0y}) = 2m_x + 2n_l m_x / \alpha_m + 2\beta_m m_x + 2n_l \beta_m m_x / \alpha_m \quad (2\text{-}24)$$

It can be deduced:

$$m_x = \frac{(g+q) l_{0x}^2}{8} \cdot \frac{1 - \dfrac{1}{3n_l}}{n_l (1 + n_l/\alpha_m + \beta_m + \beta_m n_l/\alpha_m)} \quad (2\text{-}25)$$

As the ratio of the length to the span is known, the value of m_2 can be calculated by α and β. Then the value of m'_1, m''_1, m'_2, m''_2 can be calculated out to find the required reinforcement according to the *"reinforced concrete structure design principle"* method. Similar to the elastic method, $\alpha = 1/n^2$ is also taken into account. In addition, considering the convenience of steel saving and reinforcement, β should take the value within the range of 1.5~2.5, usually $\beta = 2.5$.

Under the condition of different supporting conditions, m_1 and other corresponding bending moment values can be obtained by taking different m'_1, m''_1, m'_2, m''_2 into the Formula 2-17 according to the actual situation.

2.3.4.6 Calculation of internal force of continuous two-way slab under uniform loading

For a continuous two-way slab, suppose the variable load is full distributed. First, the middle section of the grid slab (Figure 2.20) is calculated a a single-span slab fixed on four sides. After the calculation of the middle cell, the bending moment values of the support of middle cells can be used as the known bending moments of the adjacent grid slab support. In this way, the internal forces of each section are solved sequentially from the inside to the outside. The slab cells on the edge and corner is calculated by the actual supporting condition of the boundary.

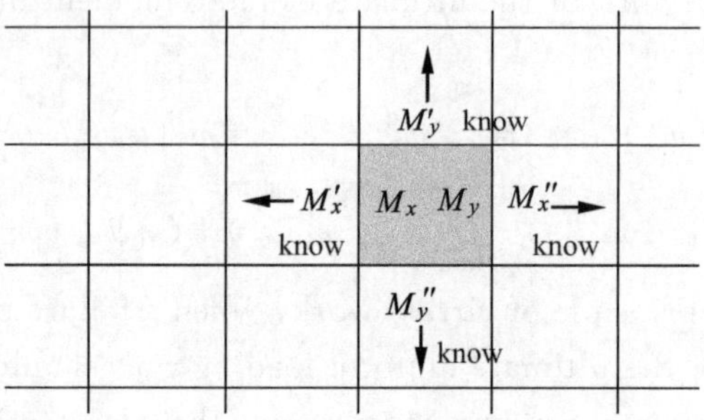

FIGURE 2.20 The calculation of internal force by plastic hinge line method for continuous two-way slab

2.3.5 Internal Force Analysis of Two-Way Slab Beam

The main difference between the two-way slab and the one-way slab is that the load on the two-way slab is transferred to both the main beam and secondary beam. The internal force calculation of two-way slab supporting beam is discussed below. (Ref. 2.6)

The load of the support beam, can be detemined by the following method: The 45-degree line angle from the four corners of each cell intersect with the center line parallel to the long sides, divide the entire board into four sections, each of which is close to the support beams (Figure 2.21). Therefore, except the self weight of beam and the load

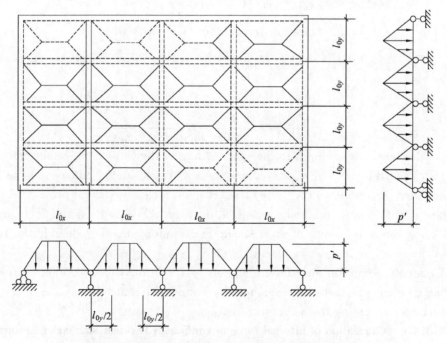

FIGURE 2.21 Load on a two-way slab supported beam

acting directly on the beam (uniform load or concentrated load), the load on the short-span beam is triangular distribution and the load on the long-span beam is trapezoidal distribution.

2.3.5.1 Calculations according to elastic theory method

For continuous support beams with equal spans or approximate spans of less than 10%, the triangle or trapezoidal load of the support beam can be converted into a uniform load. The internal forces under the uniform load of the beam can be calculated out.

Figure 2.22 shows the method of changing the triangular and trapezoidal load distribution to equivalent uniform load respectively. It is calculated according to the condition that the bending moment at the support is equal.

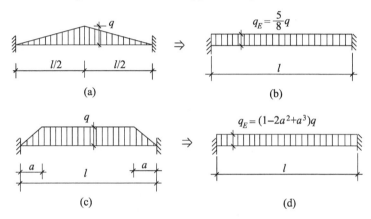

FIGURE 2.22 Equivalent uniform load

After calculating the support moment reaction, according to the equivalent uniform load (for the most unfavorable position of each live load), based on the obtained support moment and the actual load distribution of each span (triangle or trapezoidal load distribution), it can be calculated by the balance between mid-span bending moment and bearing shear force.

2.3.5.2 Calculations according to moment modulation method

When considering the redistribution of plastic internal forces, the bearing moment can be obtained according to the elastic theory. Then the bending moment is calculated by the method of amplitude modulation according to the actual load distribution moment.

2.3.6 Section Design and Construction Requirement of Two-Way Slab

2.3.6.1 Section design of slab

(1) Section's design value of bending moment

Similar to the one-way slab, if the support is a reliable constraint, the slab forms an arch (i.e. The dome effect) in both directions after cracking. Therefore, the two-way slab with the whole periphery connected with the beam should consider the influence of the surrounding support beam on the thrust of the slab. The bending moment should be reduced according to the following principle:

① The calculated bending moment of the middle span section and the middle support section can be reduced by 20%.

② Cross-section of the border area and the first inner support if:

a. $l_b/l < 1.5$, reduce by 20%;
b. $1.5 \leqslant l_b/l \leqslant 2$, reduce by 10%;
c. $l_b/l > 2$, do not reduce.

Where, l is the calculated span perpendicular to the edge of the slab; l_b is the calculated span along the edge of the slab, as shown in Figure 2.23.

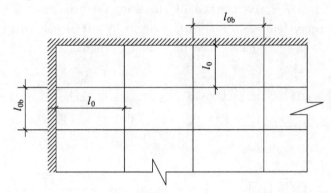

FIGURE 2.23 A schematic diagram of the calculation span of slab

(2) Effective floor height h_0

If the direction moment of short-span is larger than the direction moment of long-span, the short-span steel reinforcement should be placed outside the long-span steel bar to make full use of the effective height of the slab. Usually, the values of h_0 are as follows:

Short span direction: $h_0 = h - 20\text{mm}$ (2-26)

Long span direction: $h_0 = h - 30\text{mm}$ (2-27)

Where, h is the thickness (mm).

(3) Reinforcement calculation

The reinforcement of Two-way slab in two directions must be calculated by the design bending moment m of the unit width of the slab section. The cross-sectional area of tensile steel required is calculated by the following formula:

$$A_s = \frac{m}{\alpha_1 \gamma_s h_0 f_y} \quad (2\text{-}28)$$

where, γ_s—the arm coefficient of the internal force is approximately $0.9 \sim 0.95$.

2.3.6.2 Methods of reinforcing Two-way slabs

There are two kinds of reinforcement methods for two-way slabs: separated type and continuous type.

When the forces of a two-way slab is calculated according to the elastic theory, the mid-span moment not only changes along the length of the slab, but also decreases gradually. However, the bottom reinforcement is calculated by the maximum bending moment and should gradually reduce to both sides. For ease of construction, the slab can be divided into three strips the two directions (Figure 2.24). The width of the edge strip is 1/4 of the smaller span of the two-way slab and the rest is the intermediate slab strip. The bottom reinforcement of the middle slab is designed by the maximum bending moment obtained, the edge of the slab is reduced by half, which is not less than three per meter width. In order to withstand the torque at the four corners of the slab, the steel is uniformly distributed along the width of the support.

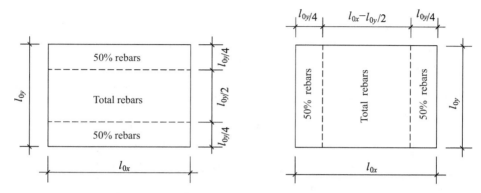

FIGURE 2.24 Calculation of the positive bending moment reinforcement of slab according to elastic theory

According to the plastic theory, the reinforcement should be consistent with the assumption of internal force calculation. The reinforcement can be evenly distributed in order to facilitate construction.

In the simply supported two-way slabs, the restraint effect of the actual bearing is not considered in the calculation, so the steel reinforcement at the bottom of the slab should be bent 1/3 in each direction.

2.3.7 Other Floor Types

The flat slab (non-beam slab) is in the slab with no ribs and directly supported by columns (Figure 2.25).

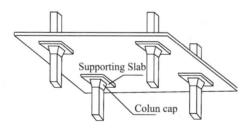

FIGURE 2.25 Flat slab with column caps and brackets

The advantages of flat slab floor are the small structural height, flat bottom, simple structure and convenient construction. According to experience, when variable load standard value of the floor is above 5kN, and the span below 6m, flat slab is more economical than ribbed slab. Therefore, flat slab is common used in factories, shopping malls, garages and other buildings.

The disadvantage of the flat slab floor structure is that the bending stiffness of the floor decreases with no rib beams and the deflection increases. The shearing stress around the column is highly concentrated, which may cause the punching failure of the local slab.

2.3.8 Platform Slab of a Workshop

The layout of the workshop's working platform is shown as Figure 2.26a. The platform slab and the beam are cast as a whole. The thickness of slab is 100mm. The slab is covered with 200mm thick cement sand plaster. The strength grade of concrete is C20 and the steel bar is HPB235 grade. Live load on board is $q=4\text{kN/m}^2$. Get the internal moment of the workshop's platform slab by elastic theory.

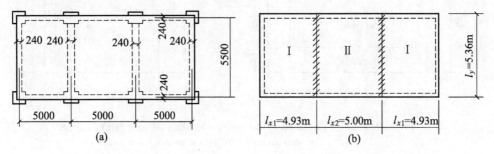

FIGURE 2.26 The layout of workshop's working platform

2.3.8.1 Calculating diagram

The calculating diagram is shown as Figure 2.26b.

Calculation span is

$$l_{x1} = l_{轴} + \frac{h}{2} = (5000-120) + \frac{100}{2} = 4930\text{mm}$$

$$1.05l_0 = 1.05 \times (5000-240) = 4998\text{mm}$$

$$l_{x1} = 4.93\text{m}, l_{x2} = l_{轴} = 5.0\text{m}$$

$$l_y = l_0 + 2 \times \frac{h}{2} = (5500-240) + 2 \times \frac{100}{2} = 5360\text{mm} = 5.36\text{m}$$

2.3.8.2 Load calculation

Standard value of constant load:

slab weight	$0.1 \times 25 = 2.50\text{kN/m}^2$
surface weight	$0.02 \times 20 = 0.40\text{kN/m}^2$

The platform is controlled by the combination of variable load effect:

design value of constant load	$g = 2.9 \times 1.2 = 3.48\text{kN/m}^2$
design value of live load	$q = 4.0 \times 1.4 = 5.60\text{kN/m}^2$
design value of total load	$g+q = 3.48 \times 5.60 = 9.08\text{kN/m}^2$

2.3.8.3 Internal force calculation

(1) Positive bending moment of mid-span

$$g + \frac{q}{2} = 3.48 + \frac{5.60}{2} = 6.28\text{kN/m}^2$$

$$\frac{q}{2} = \frac{5.60}{2} = 2.80\text{kN/m}^2$$

Grid I: $\dfrac{l_x}{l_y} = \dfrac{4.93}{5.36} = 0.92$

From Figure 2.27, every one meter wide strip is taken in two directions.

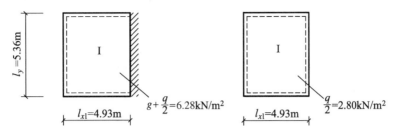

FIGURE 2.27 The positive bending moment of grid I

Check the appendix B-2

$$m_{x1} = \text{Table's coefficient} \times \left(g+\frac{q}{2}\right)l_{x1}^2 = 0.0382 \times 6.28 \times 4.93^2 = 5.83 \text{kN} \cdot \text{m}$$

$$m_{y1} = \text{Table's coefficient} \times \left(g+\frac{q}{2}\right)l_{x1}^2 = 0.0234 \times 6.28 \times 4.93^2 = 3.57 \text{kN} \cdot \text{m}$$

Check the appendix B-1

$$m_{x2} = \text{Table's coefficient} \times \frac{q}{2}l_{x1}^2 = 0.428 \times 2.80 \times 4.93^2 = 2.91 \text{kN} \cdot \text{m}$$

$$m_{y2} = \text{Table's coefficient} \times \frac{q}{2}l_{x1}^2 = 0.362 \times 2.80 \times 4.93^2 = 2.46 \text{kN} \cdot \text{m}$$

The superposition of the two value is the positive bending moment of the middle span in grid I

$$m_x = m_{x1} + m_{x2} = 5.83 + 2.91 = 8.74 \text{kN} \cdot \text{m}$$
$$m_y = m_{y1} + m_{y2} = 3.57 + 2.46 = 6.03 \text{kN} \cdot \text{m}$$

Grid II: $\dfrac{l_x}{l_y} = \dfrac{5.0}{5.36} = 0.93$

From Figure 2.28, every one meter wide strip is taken in two directions.

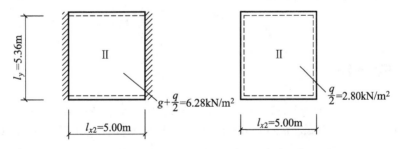

FIGURE 2.28 The positive bending moment of grid II

Check the appendix B-3

$$m_{x1} = \text{Table's coefficient} \times \left(g+\frac{q}{2}\right)l_{x1}^2 = 0.0305 \times 6.28 \times 5^2 = 4.79 \text{kN} \cdot \text{m}$$

$$m_{y1} = \text{Table's coefficient} \times \left(g+\frac{q}{2}\right)l_{x1}^2 = 0.0143 \times 6.28 \times 5^2 = 2.25 \text{kN} \cdot \text{m}$$

Check the appendix B-1

$$m_{x2} = \text{Table's coefficient} \times \frac{q}{2}l_{x2}^2 = 0.0419 \times 2.80 \times 5^2 = 2.93 \text{kN} \cdot \text{m}$$

$$m_{y2} = \text{Table's coefficient} \times \frac{q}{2}l_{x2}^2 = 0.0363 \times 2.80 \times 5^2 = 2.54 \text{kN} \cdot \text{m}$$

The superposition of the two value is the positive bending moment of the middle span in grid Ⅱ

$$m_x = m_{x1} + m_{x2} = 4.79 + 2.93 = 7.72 \text{kN} \cdot \text{m}$$
$$m_y = m_{y1} + m_{y2} = 2.25 + 2.54 = 4.79 \text{kN} \cdot \text{m}$$

(2) Negative bending moment of support

Grid Ⅰ: $\frac{l_x}{l_y} = 0.93$

From Figure 2.29, every one meter wide strip is taken in two directions, check the appendix B-2:

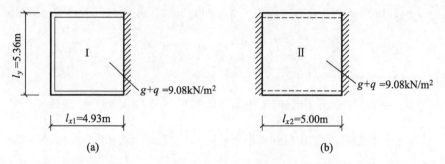

FIGURE 2.29 The positive bending moment of support

$$m'_{x1} = \text{Table's coefficient} \times (g+q)l_{x1}^2 = 0.0897 \times 9.08 \times 4.93^2 = 19.80 \text{kN} \cdot \text{m}$$

Grid Ⅱ: $\frac{l_x}{l_y} = 0.94$

From Figure 2.29b, every one meter wide strip is taken in two directions, check the appendix B-3: $m'_{x2} = \text{Table's coefficient} \times (g+q)l_{x2}^2 = 0.0725 \times 9.08 \times 5^2 = 16.46 \text{kN} \cdot \text{m}$

Take the mean value of the two:

$$m'_x = \frac{m'_{x1} + m'_{x2}}{2} = \frac{19.8 + 16.46}{2} = 18.13 \text{kN} \cdot \text{m}$$

2.4 Stairs

Stairs are the vertical transportation for multi-storey buildings and are the main evacuation routes for high-rise buildings in case of fire and other disasters. The common stairs are mainly cast-in-place of reinforced concrete, and the reinforced concrete precast stairs are also used in multi-storey buildings in non-seismic areas. The structure of stairs are mainly slab-beam style stairs. In some public buildings, scissors and spiral stairs are widely used.

This section mainly introduces two basic stairs: slab stairs and beam stairs (Figure 2.30).

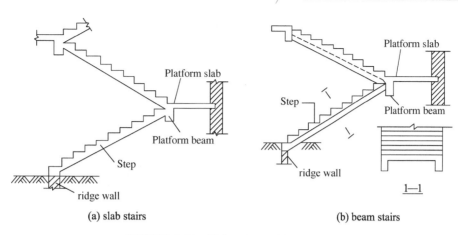

FIGURE 2.30 Slab stairs and beam stairs

2.4.1 Slab Stairs

The typical slab components include slant slab, platform slab and platform beam (Figure 2.31). Both ends of the oblique slab are supported by floor beams. Its advantages are the even face and convenient construction of the formwork. The disadvantage is that when the span is larger, the sloping slab is thick and more material is used. Therefore, the slab stair is suitable for the cases with small variable loads and for the spans not greater than 3m.

2.4.2 Beam Stairs

Cast-in-place beam staircase includes the ladder, platform slab and platform beams. The ladder consists of a slab and two inclined beams. The load is transmitted from the slab to the inclined beams, and then to the platform beams, and finally transmitted to the wall or columns. When the span is greater than 3m, it is more economical to use the beam staircase. The disadvantage of the beam staircase is that the construction of the supporting formwork is more complex.

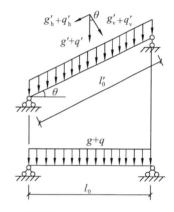

FIGURE 2.31 Slab stairs

When calculating slab, strip with 1m width is taken as calculating unit. The maximum bending moment in the span of ladder slab can be calculated out:

$$M_{max} = \frac{1}{8}(g'_v + q'_v)l'^2_0 = \frac{1}{8}(g+q)\cos\theta \cos\theta \frac{l_0^2}{\cos^2\theta} = \frac{1}{8}(g+q)l_0^2 \quad (2\text{-}29)$$

where, l'_0 is oblique span of slab, l_0 is horizontal span of slab, θ is the angle between slab and horizontal plane.

The mid-span bending moment of beam staircase pedal is:

$$M = \frac{1}{8}(g+q)l_0^2 \quad (2\text{-}30)$$

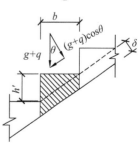

Figure 2.32 Calculation sketch of beam staircase

When the two ends of the pedal are connected as a whole with the inclined beam of the staircase section, the mid-span bending moment of the pedal is:

$$M = \frac{1}{10}(g+q)l_0^2 \tag{2-31}$$

References

2.1　Gu XiangLin. Building concrete structure design[M]. Shanghai: Tongji University Press, 2011.

2.2　Xu Jinsheng, Xue Lihong. Modern prestressed concrete floor structure[M]. Beijing: China Construction Industry Press, 1998.

2.3　Honestad E. Yield line theory for the ultimate flexural strength of reinforced concrete slabs[J]. J. ACI, 1953: 637−656.

2.4　Code for design of concrete structures[S]. GB 50010−2010. Beijing: China Architecture and Building Press, 2010.

2.5　Nichols J R. Statical limitations upon the steel requirement in reinforcement in reinforced concrete flat slab floors[J]. Trans. ASCE, 1914: 1670−1736.

2.6　Gamble W L, Sozen M A, Siess C P. Test of a two-way reinforced concrete floor slab[J]. J. Strcut. Div,. ASCE, 1969: 1073−1096.

Questions

2.1　What is plastic hinge? What is a plastic hinge line? What is the redistribution of plastic internal forces?

2.2　What is the difference between one-way slab and two-way slab?

2.3　What are the advantages and disadvantages of the plate staircase and the beam staircase?

2.4　Design the cast-in-place staircase of a teaching building. The plan and section size of the stair is shown in Figure 2.33. The strength grade of concrete is C20 and the grade of steel bar is HPB235. Floor's live load is $q_k = 2.5 \text{kN/m}^2$.

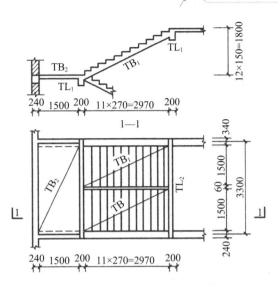

FIGURE 2.33 Stair's plan and section size

2.5 As shown in Figure 2.34, the side length is 5m, the slab thickness is 100mm, the concrete degree is C30, the up and bottom reinforcement are: $\phi 8 @ 200$, the reinforcement at the support is $\phi 8 @ 150$. Please calculate the slab's ultimate uniform load by the elastic method and plastic hinge line method respectively.

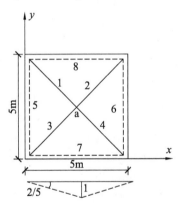

FIGURE 2.34 The two-way slab

Chapter 3
Single-Storey Factory Building

3.1 Introduction

Industrial factory buildings can be divided into single-storey and multi-storey buildings according to the number of layers. Metallurgy and machinery plant, such as forging, welding, metalworking, assembly and machine repair shops, often have heavy equipments. The machines in these factories are often heavy and large in shape, which is difficultly to be moved and increases the load of the structures. Thus, single storey factory building is a best choice. (Ref. 3.1).

According to the scale of production, single-storey factory buildings can be divided into large, medium and small factories.

In general, a small factory building is constructed with no crane or crane tonnage is not more than 5t, Its span is less than 15m, and the top elevation of the column is below 8m, no special process requirements is involved and a hybrid structare system including brick column, reinforced concrete roof beam or light steel roof trusses is often used. For large factories, when the tonnage of crane is above 250t, the span is larger than 26m, special technological requirements is required and all-steel structure or reinforced concrete column and steel roof truss are often used. The other factory buildings can be made of reinforced concrete structures, precast and prestressed concrete members.

According to structural form, single storey factories can be divided into two main kinds: portal frame structures and bent frame structures. For portal frame structures, as shown in Figure 3.1, the beam-column joints are rigid and the columns are hinged to the foundation. For bent frame structures, as shown in Figure 3.2, the beam-column joints are hinged and the columns are fixed to the foundation.

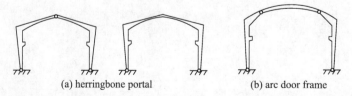

(a) herringbone portal   (b) arc door frame

FIGURE 3.1 Portal frame structure system of single story factory building

Bent frame structures have different forms, such as equal height form (Figure 3.2a), unequal height form (Figure 3.2b) and zigzag form (Figure 3.3). The bent frame structure has the advantages of definite force transmission, simple structure, convenient stereotype design, standardized structural parts, industrialized production and facilitation of mechanized construction.

This chapter mainly deals with the main problems in the design of prefabricated reinforced concrete bent frame structures.

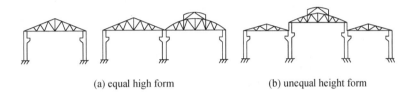

(a) equal high form   (b) unequal height form

FIGURE 3.2   High and unequal height bent structures

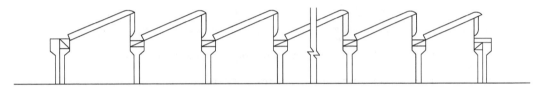

FIGURE 3.3   Zigzag bent frame structure

3.2   Structural System and Design of Single-Layer Factory Buildings

3.2.1   Structure System

The single-storey bent frame structure are composed of the main load-bearing components and the secondary components. The main load-bearing components include the roof truss (or roof beam), column, crane beam and foundation. The secondary components are roof panel, strut, lintel, etc.. (Figure 3.4) These components constitute a space structure, which can be divided into three force stress systems (Ref. 3.2).

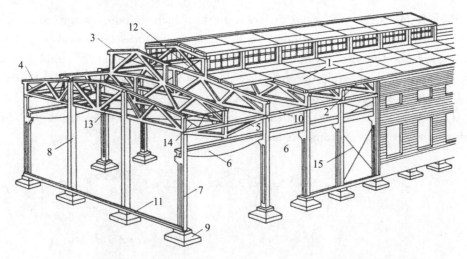

1—roof panel; 2—gutter board; 3—skylight frame; 4—roof truss; 5—bracket; 6—crane beam; 7—bent column; 8—wind resistant column; 9—foundation; 10—tie beam; 11—foundation beam; 12—skylight frame vertical bracing; 13—lateral bracing at the bottom of the roof truss; 14—vertical bracing at the end of the roof truss; 15—column bracing

FIGURE 3.4 Structural Composition of Single-Storey Factory Building

3.2.1.1 Enclosure structure system roof and wall

The roof structure is divided into purlin system and no purlin system. The No purlin system consists of large roof slab, roof girder and roof truss (including roof bracing), the roof panels and roof trusses are welded together, so that the structure has good integrity and stiffness. This system is suitable for large and medium-sized single-storey factory buildings because it has few components and its construction speed is fast. The Purlin system consists of small roof panel, purlin and roof truss (including roof bracing), its roof structure is small in size and light in weight, so its transportation and installation is easier. This system is often used in non-insulated small workshops.

The main functions of the roof structure are acting as enclosure and bearing load (bearing the weight of the roof structure, live load on the roof, snow load and other loads), as well as lighting and ventilation. Sometimes, the roof structure is also provided with a skylight frame setting for ventilation and lighting.

The wall enclosure structure includes exterior wall, wind resistant column, wall beam and foundation beam, etc. Its main functions is bearing the weight of the wall and other components, and the wind load acting on the wall.

3.2.1.2 Transverse bent frame structure system

The transverse plane bent frame consists of roof beam (or truss), transverse column and foundation (Figure 3.5), it is the basic bearing structure commonly used. The vertical loads including the weight of the structure, the live load of the roof, the snow load and the vertical load of the crane, are supported by the transverse bent frame. The horizontal load sustained by the structure includes the wind load, the lateral braking force of the crane and the earthquake action. The load transfer path on the transverse bent frame is shown in Figure 3.6a and b.

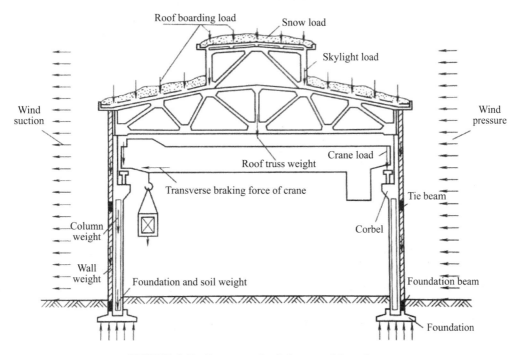

FIGURE 3.5 Transverse load diagram of bent frame

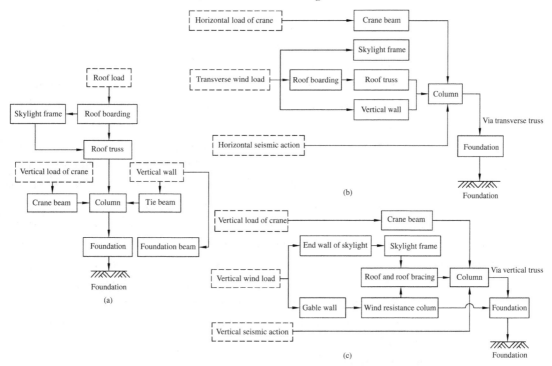

FIGURE 3.6 Main load transfer paths on the transverse bent

3.2.1.3 Longitudinal bent frame structure system

The longitudinal plane bent includes column, foundation, continuous beam, crane beam and column bracing (Figure 3.7). Its function is to ensure the longitudinal stability and load-bearing of the structure. It mainly resists the longitudinal horizontal load, such as longitudinal wind load, longitudinal braking force of the crane, longitudinal seismic action and thermal stress. The main load transfer paths of the longitudinal bent frame structure are shown in Figure 3.7.

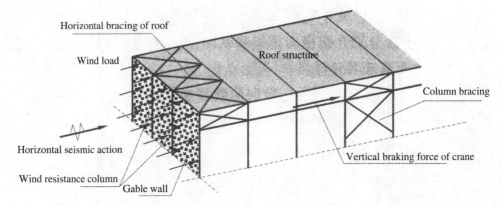

FIGURE 3.7 Loads on the longitudinal plane bent

3.2.2 Layout of Structure Members

3.2.2.1 Laynot of column

The distance between the longitudinal axis of the bearing column (or bearing wall) of the plant is called the column span. The distance between the horizontal axes is called the column pitch. Column layout is used to determine the column position, span of roof panels, roof truss (or roof beam), crane beam, etc..

In order to facilitate the design of the plant structure, the production and construction of the components, and ensure the standardization and finalization of the structural components, the column size should be consistent with the unified modular system as specified in the Harmonized Modular Building Standards (GB / T 50006—2010) taking 100mm as the basic unit, expressed with "M". When the plant span is not more than 18m, a multiple of 3m (30M), i.e. 9m, 12m, 15m and 18m should be adopted; when the span of the plant is greater than 18m, the horizontal span of the plant shall be a multiple of 6m, if necessary, also allows the use of 21m, 27m, 33m span. Longitudinal column spacing plant generally use 6m or multiples of 6m (60M), individuals may also use 9m or other column spacing.

The main plane dimensions of the structure are represented by axes, the axis direction of the span is called the axis of transverse positioning, expressed with No. ①, ②,③, ... , the axis direction of the column spacing is called longitudinal positioning, expressed with No. A, B,C ...

3.2.2.2 Deformation joints

Deformation joints include expansion joints, settlement joints and seismic joints.

(1) Expansion joints

If the length or width of a single-storey factory is too large, with the change of temperature, the upper part of the plant will expand with heat and contract with cold, but the underground part will not be affected greatly by the temperature. As a result, the superstructure of the plant exposed to the atmosphere is restricted in its extension and contraction and temperature stresses generate inside the structure (i. e. columns, walls, longitudinal crane girders and tie beams). When the temperature stress becomes larger, the roof and the walls may develop cracks, which reduce the bearing capacity of the column and affect the normal use of the plant. In order to reduce the adverse effects of temperature changes on the plant, expansion joints along the plant's horizontal and vertical directions (Figure 3.8) are required to divide the plant into several temperature sections. The division of the temperature section should be as simple and regular with minimum the number of expansion joints.

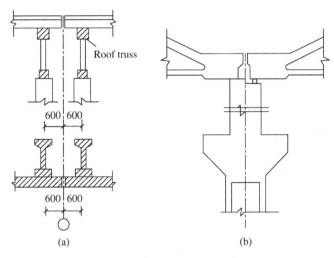

FIGURE 3.8 Installation of expansion joint

(2) Settlement joints

When the height or the crane lifting weight difference between adjacent buildings is large, the compressibility of the foundation and the structural type of the plant is obviously different, a settlement joint should be set up. The settlement joint separates the plant structures into several individual parts (including the foundation) and it can also be used as expansion joints.

(3) Seismic joints

Seismic joints are one of the measures to reduce the seismic damage of the factory building. When the factory buildings is complex or has attached buildings, anti-vibration joints should be provided. The width of seismic joints can be 100~150mm at the vertical and horizontal junction when the plants has a large column grid or no column bracings, otherwise 50~90mm can be used. At least one side of the transition span of the two buildings should be separated from the main plant by anti-vibration joints.

3.2.2.3 Plant height

The height of the plant is generally expressed by the truss bottom elevation H_1 and the crane rail top elevation H_2. (Figure 3.9). For plant with no crane, the truss bottom elevation is determined by the height of the machine and the production requirements of the factory. For the crane truss, the bottom elevation is determined by following formula:

$$H_1 = \max \begin{Bmatrix} h_1+h_2+h_3+h_4+h_5+h_6+h_7 \\ h_1+h_2+h_8+h_5+h_6+h_7 \end{Bmatrix} \quad (3\text{-}1)$$

where, h_1—the height of the highest equipment in the plant, determined by the production requirements. If the equipment is too high, it may be partially buried underground to reduce the h_1.

h_2—safe height, generally not less than 500mm.

h_3—heavy lifting height.

h_4—minimum sling height.

h_5—the height from the crane bottom to the top of the crane rail.

h_6—the height from the top of the crane rail to the top of the crane.

h_7—safe height of crane running, generally not less than 220mm.

h_8—the height from the bottom of the driver's cab to the bottom of the crane, refer to the relevant crane product manual.

Rail top elevation can be determined by following formula:

$$H_2 = H_1 - h_6 - h_7 \quad (3\text{-}2)$$

The Uniform Modulus requirement should be taken into account when determining H_1 and H_2.

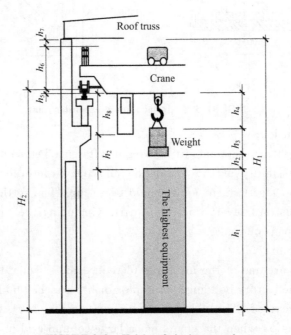

FIGURE 3.9 Height of single span plant

3.2.2.4 Strut layout

Single-storey factory building is an assembly structure which is composed of prefabricated components connected and assembled on site. The layout of the bracing is an important part of the structural design of a single-storied building. Improper support arrangement will affect the normal use of the factory building, even lead to engineering quality accidents.

The bracing of a single-storey factory building with bent structure can be categorized into two: Roof bracing and Column bracing. The cross-sectional design of each bracing member is generally determined by the structural requirements. The following is a brief description of the layout principles of these two types of bracing in non-seismic areas.

1) Roof bracing

Roof bracing system includes horizontal bracing, longitudinal horizontal bracing, vertical bracing and tie bar.

(1) Lateral horizontal bracing

The lateral horizontal bracing system consists of cross angle steel and the upper or lower chords of horizontal truss (Namely upper clord and lower chord lateral horizontal bracing respectively) (Figure 3.10). The lateral horizontal bracing is generally arranged at both ends of the plants temperature section. Its help strengthen the rigidity of the roof within the horizontal level, and also transmit the loads from the gable or wind columns to the columns on both sides.

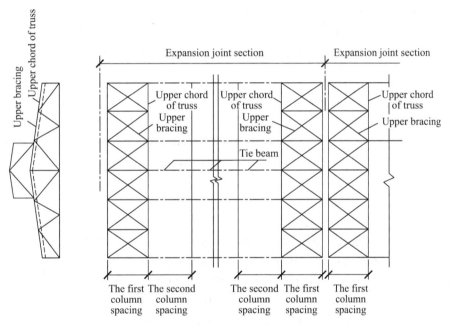

FIGURE 3.10　Upper lateral bracing

(2) Longitudinal horizontal bracing

Longitudinal horizontal bracing system consists of cross angle steel and lower cord of roof truss. It is usually set at the first and middle internodes of the roof truss (Figure 3.11). The role of longitudinal horizontal bracing is to enhance the lateral horizontal stiffness of the roof.

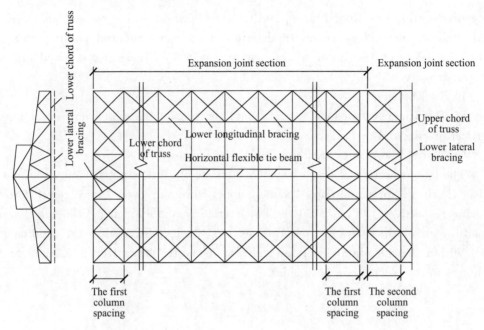

FIGURE 3.11 Lower lateral bracing and lower longitudinal bracing

(3) Vertical bracing

The vertical bracing system are usually composed of vertical truss of cross section angle steel and vertical web of the roof truss or column of the skylight frame. Vertical bracing are typically located at the ends or mid-span of the truss of the plant temperature section. Its main role is to ensure the lateral stability of the roof truss, and used to transfer the longitudinal horizontal force. Therefore, the vertical bracing should be used with lateral horizontal bracing.

(4) Tie bar

Tie bar is a single connection member. Ties that only withstand tensile stress are called flexible tie bars, usually steel bars. Tie bars that can resist both tensile and compressive stree are called rigid tie bar, which can be steel bars or reinforced concrete bar members. Tie bar is usually connected to the vertical bracing or the nodes of the upper and lower chord of lateral horizontal bracing. The function of the tie rod is acting as a lateral fulcrum of the roof truss or roof girder to ensure the global stability of the roof truss and the lateral stability of the upper roof truss or compressive flange of roof beam, avoid excessive lateral vibration generated by working crane in the plant.

2) Inter-column bracings

The horizontal load on the gable is transmitted to the columns on both sides of the factory building through the roof and bracing system. When the crane moves longitudinally along the factory floor, the generated horizontal load is also be transmitted to the column. In order to ensure that the longitudinal column of the plant can work together and resist the horizontal forces, inter-column bracings are usually used along the longitudinal columns of the plant. The bracing set above the corbel is called upper inter-column bracing, and that set below the corbel is called lower inter-column bracing.

Inter-column bracings used are generally crossed steel diagonal rod. The upper inter-column bracings generally arranged at the two ends of the temperature sections corresponding to the lateral horizontal bracings of the roof, and the center of the temperature section. The lower inter-column bracings are generally arranged at the center of the temperature section and the corresponding position between the upper column bracings (Figure 3. 12).

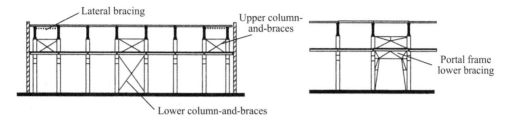

FIGURE 3. 12 Arrange of the column-and-braces

3.2.2.5 Layont of exterior protected construction

(1) Wind column

The gable wall of the plant has larger wind area, which is usually divided into several segments by the wind column, and the wind load on the wall (the section near the longitudinal column) is directly transmitted to the longitudinal columns. Part of the load transfers the foundation and part transfers to the roof frame system via the top wind column. (Figure. 3. 13).

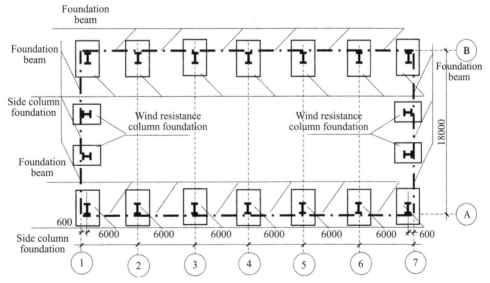

FIGURE 3. 13 The foundation layout of the single-storey factory building

(2) Ring beam, tie beam, lintel and foundation beam

Ring beam, tie beam, lintel and foundation beam will be set when the brick wall works as retaining wall.

The purpose of setting the ring beam is to surround the wall, frame column and wind column, increase the overall rigidity of the building and prevent the adverse effects caused by uneven settlement of the foundation or larger vibration load. Tie beam only play a tie role, they don't bear the weight of the wall, so it is unnecessary to set a corbel that brace the tie beam on the column.

In addition to support the wall weight, the tie beam can also link the longitudinal columns to enhance the longitudinal stiffness of the plant and transfer longitudinal load. The two ends of the tie beam are supported on the column corbels and can be bolted or welded.

Lintel beams is generally precast reinforced concrete components, whose role is to support the weight of the wall on the upper part of a door hole or window openings. The width of the section is generally the same as the thickness of the wall.

When the masonry strength of a single-storey factory building is sufficient to withstand the weight of a wall, a foundation beam is generally used to support the retaining wall, and the weight of the wall is directly transferred to the plinth without the use of a wall foundation (Figure 3.14).

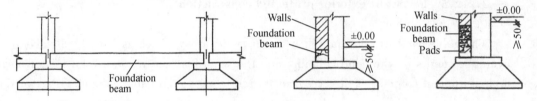

Figure 3.14 Foundation beam

3.3 Internal Force Calculation of Bent Frame Structure

3.3.1 Assumptions and Calculations

The vertical load of a single story factory is mainly transmitted through transverse bent frame. Therefore, the calculation and analysis of bent frames are mainly based on transverse bent frame. According to the construction of the single story bent structure plant, the following simplifications and assumptions are made to determine the calculation diagram(Ref. 3.3).

(1) Take a transvers bent frame as the basic unit of calculation (Figure 3.15a).

(2) The ends of the roof and top of upper columns are usually welded by embedded steel plates, but its ability to resist bending moment is very small and can be simplified to a hinge connection (Figure 3.15c).

(3) When the bent column is inserted into a certain depth of the foundation, it is generally poured with high-grade fine stone concrete and fixed by the foundation (Figure 3.15b).

(4) The rigidity of roof beam or roof truss is very large and the axial deformation can be neglected subjected to force, which can be simplified as an infinite-rigid body. It

is not suitable for composite roof truss due to its relative small stiffness, and should be carefully considered.

(5) The height from the top level of the base to the top of the column is called bent column height. The axis of a column is the geometric center line of the column.

(6) The span of the bent frame is based on the axis of the plant.

The final calculation sketch of the transverse bent frame is shown in Figure 3.15d.

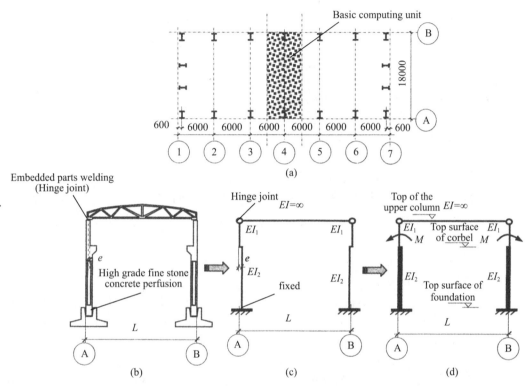

FIGURE 3.15 Calculation sketch of transverse bent

3.3.2 Standard Load Values of Bent Frame Structure

3.3.2.1 Roof load

The roof load P_1 is transmitted to the top of bent column by the contact between the roof truss and the upper column. When there is eccentricity, additional bending moments are added to the calculation as shown in the diagram (Figure 3.16).

The permanent load of the roof load mainly includes the self-weights of the roof structural layer, roof slab, skylight frame, roof truss and the roof brace system which can be calculated according to the material density and size of the components. The variable loads of the roof load mainly include snow load, roof load and dust accumulation load.

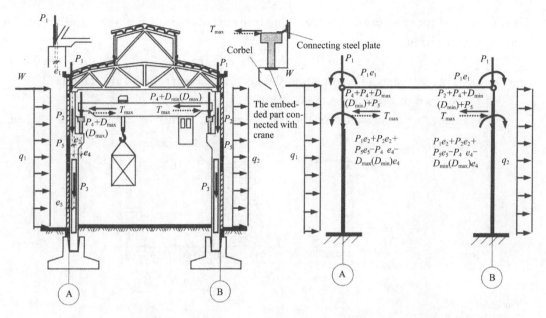

FIGURE 3.16 Load standard values of horizontal bent

The standard value of snow load on the horizontal projection surface of the roof is calculated by:

$$S_k = \mu_r S_0 \tag{3-3}$$

where, S_k—Snow load, kN/m^2;

S_0—Basic snow pressure, kN/m^2, is determined by the dead weight of the largest snow recorded in 50 years on a flat, open surface in the Local area. For Details refer to the "*Load Code for the Design of Building Structures*" (GB 50009)(Ref. 3.4)

μ_r—Snow distribution factor of roof. When the roof slope is less than 25°, $\mu_r = 1.0$.

Live load on roof refers to the load during the construction or maintenance on the ummanned roof. Roof live load and snow load are considered simultaneously, however, the larger value should be taken.

Roof ash loads are considered in the roof of the machinery, metallurgy and cement plant. The values of the roof ash load of the roof area can be found in *Load Code for the Design of Building Structures*(GB 50009)(Ref. 3.4).

3.3.2.2 Column weight and wall weight

The weight of the upper column P_2 and the weight of the lower column P_3 can be calculated according to the bulk density of the material and the size of the component. At the same time, the bending moment $P_2 e_2$ acting on the lower column and caused by weight of the upper column P_2 should also be considered.

If there is a wall beam in the wall and the wall beam is connected with the bent frame, the wall weight P_5 transmitted from the wall beam and its corresponding bending moment $P_5 e_5$ should be taken into account (Figure 3.16).

3.3.2.3 Load from crane beam

Load from crane beam includes weight of crane beam and track P_4, vertical load $D_{max}(D_{min})$ acting on the top of the corbel which is caused by the vertical wheel pressure of the crane, and the horizontal force T_{min} produced on the bent column during the horizontal braking of the crane (Figure 3.17).

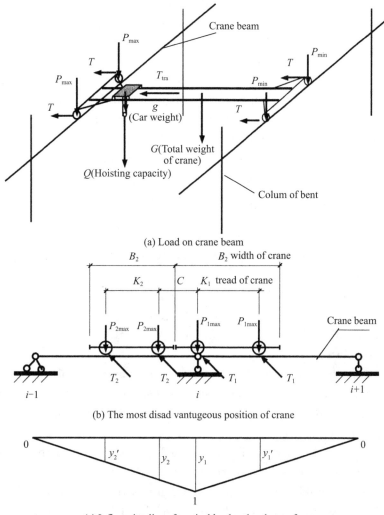

FIGURE 3.17 Load of bridge crane

The weight P_4 can be calculated according to the bulk density and the size of the component.

The bridge crane runs on the crane beams, and its load acts on the crane beams as the wheel pressure of the crane. When the crane is fully loaded and the small car reaches the limit position of one end of the bridge, the maximum wheel pressure produced near the end of the car is P_{max}, and the minimum wheel pressure near the other end is P_{min} (Figure 3.17a). For the four wheel crane, the relationship between P_{max} and P_{min} is:

$$P_{max}+P_{min}=\frac{G+Q}{2} \tag{3-4}$$

where, G—total weight of crane;
Q—lifting weight of crane.

The wheel pressure of the crane is transmitted to the corbel through the crane beam when the crane is moving. Therefore, it is necessary to use the concept of influence line to calculate the vertical load of corbel. The crane beam is welded by the embedded iron parts at the underside of the beam and on the top of corbel (Figure 3.15). Thus, the crane beam under vertical load is simply supported on the corbel (Figure 3.17b). Assuming that the simple support shown in Figure. 3.17b is the corbel shown in the calculation diagram (Figure 3.15), the influence line of the vertical load of the corbel (or the force of the bearing) can be drawn in Figure 3.17c when the concentrated load acts on the crane. If there are more than one bridge cranes in the workshop, the vertical load acting on the corbel surface can be obtained according to the most unfavorable position of the crane shown in Figure 3.17b and the influence line shown in Figure 3.17c. The formula is as following:

$$\begin{cases} D_{max} = \sum P_{imax}(y_i + y_i') \\ D_{mix} = \sum P_{imix}(y_i + y_i') \end{cases} \quad (3\text{-}5)$$

Based on D_{max} and D_{min}, the corresponding bending moment acting on the bent frame can be calculated out:

$$\begin{cases} M_{max} = D_{max} e_4 \\ M_{mix} = D_{mix} e_4 \end{cases} \quad (3\text{-}6)$$

When calculating D_{max}, D_{min}, M_{max} and M_{min}, attention must be given to the following two points:

Considering the possibility of simultaneous operation of multiple cranes, a maximum of two cranes are considered simtuniously for single-span bent frame factory buildings, and a maximum of four cranes are considered simtuniously for a bent frame of multi-span factory buildings.

D_{max} may be acted either on the A axis in Figure 3.16 or on the B axis in Figure 3.16, the above two cases should be considered.

The horizontal braking force T_{tra} of the crane is the friction caused by starting or braking of the car when it moves with lifting weight. For soft hook crane, T_{tra} shall be calculated in accordance with the following formula:

$$\begin{cases} T_{tra} = 0.12(Q+g) & Q \leqslant 10t(98kN) \\ T_{tra} = 0.10(Q+g) & Q = 15 \sim 50t(147 \sim 490kN) \\ T_{tra} = 0.08(Q+g) & Q \geqslant 75t(735kN) \end{cases} \quad (3\text{-}7)$$

where, g—weight of car;

Q—lifting weight of crane.

For the hard hook crane, it can be calculated according to the following formula:

$$T_{tra} = 0.12(Q+g) \quad (3\text{-}8)$$

The wheel of the bridge crane large car delivers T_{tra} to the track of the crane beam. Suppose the horizontal forces T transmitted by each wheel and passed to the crane beam are equal.

$$T = \frac{T_{tra}}{n_w} \quad (3\text{-}9)$$

where, n_w—the number of wheels for a crane.

Crane beams near the top of the column and the corresponding parts of the upper column of the corbel column (the top of the corbel) are embedded with iron pieces. The connecting steel plate is welded between the embedded iron pieces of the crane beam and the upper column. The horizontal force acting on the top of the crane beam is transmitted to the upper column of the corbel column through the connecting steel plate, as shown in Figure 3.16. According to the connection structure between the crane beam and the bent column, the crane beam under the horizontal load is considered to be simply supported on the bent frame. The most unfavorable position of the horizontal load of the crane and the influence line of the horizontal load of the bent column are shown in Figure 3.17b and c. According to the influence line of Figure 3.17, when multiple cranes are working at the same time, the horizontal load T_{max} at the top surface of the crane beam can be calculated.

$$T_{max} = \sum T_i(y_i + y_i') \tag{3-10}$$

The following two points should be considered when calculating T_{max}:

(1) There are two possible directions of T_{max}.

(2) No matter single span or multi span workshops, only two cranes are considered for each row of bents.

Considering the low possibility that multiple cranes suffer extreme loads at the same time, the load reduction factor should be multiplied for D_{max}, D_{min} or T_{max} achieved. According to Formula (3-5) and Formula (3-10), we can decrease the value by multiplied by a reduction factor β_D, as shown in Table 3.1.

TABLE 3.1 Load reduction factor of multiple cranes

Number of cranes participating in combination	Crane working level	
	A1~A5	A6~A8
2	0.90	0.95
3	0.85	0.90
4	0.80	0.85

EXAMPLE 3.1

A single-layer one-span workshop has a span of 24m with column spacing 6m. Two bridge crane with soft hook and working levels of A3, 20/3t (lifting crane weight of the size hook) is considered. The small car weight is 70.1kN, the maximum width of crane B is 5160mm, the maximum wheel pressure P_{max} is 183kN, the minimum wheel pressure P_{min} is 40kN, crane wheel distance K is 4100mm. Determine D_{max}, D_{min} and T_{max} on the bent frame.

SOLUTION

The influence line of the bearing reaction force is shown in Figure 3.18:

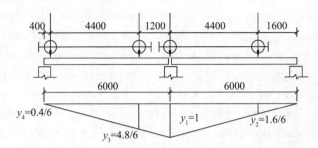

FIGURE 3.18 The influence line of the bearing reaction force

$$D_{max} = \sum P_{imax}(y_i + y'_i) = 183 \times \left(1 + \frac{4.94}{6} + \frac{1.9}{6} + \frac{0.84}{6}\right) = 417.24 \text{kN}$$

$$D_{min} = \sum P_{imin}(y_i + y'_i) = 40 \times \left(1 + \frac{4.94}{6} + \frac{1.9}{6} + \frac{0.84}{6}\right) = 91.2 \text{kN}$$

Because the crane is a soft hook crane and its weight is greater than 10t, the horizontal braking force of the crane T_{tra} is calculated by:

$$T_{tra} = 0.12(Q+g) = 0.10 \times (196 + 29.4) = 22.54 \text{kN}$$

The horizontal force transmitted by each wheel to the crane beam is as follows:

$$T = \frac{T_{tra}}{n_w} = \frac{22.54}{4} = 5.64 \text{kN}$$

The horizontal load at the top of the column crane beam on the bent frame is:

$$T_{max} = \sum T_i(y_i + y'_i) = 5.64 \times \left(1 + \frac{4.94}{6} + \frac{1.9}{6} + \frac{0.84}{6}\right) = 12.86 \text{kN}$$

As the cranes working level is A3, the reduction factor 0.9 should be multiplied according to Table 3.1.

$$D_{max} = 417.24 \times 0.9 = 375.52 \text{kN}$$
$$D_{min} = 91.2 \times 0.9 = 82.08 \text{kN}$$
$$T_{max} = 12.86 \times 0.9 = 11.6 \text{kN}$$

3.3.2.4 Wind load

The calculation of wind load has been described in the first chapter. Here is an example of bent frame structure.

EXAMPLE 3.2

A metalwork shop size and its wind load shape coefficient is shown in Figure 3.19, the basic wind pressure ω_0 is 0.35kN/m², the height of the column is +10.5m, the elevation of outdoor is −0.30m, the ground roughness class is B, and the width of bent frame B is 6m. Determine the design value of wind load acting on bent frame.

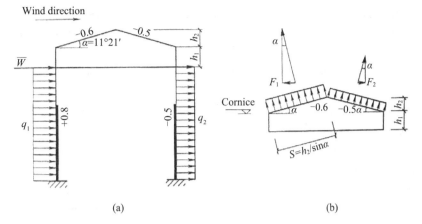

FIGURE 3.19 The workshop size and wind load shape coefficient

SOLUTION

(1) The height from the top of the column to the elevation of outdoor:
$$H = 10.5 + 0.3 = 10.8\text{m}$$
According to Table 1-4 the variation coefficient of wind pressure height is:
$$\mu_z = 1 + \frac{1.13 - 1.00}{15 - 10} \times (10.8 - 10) = 1.02$$

It is safe to adopt 1.02 for the wind pressure variation coefficient along the column:
$$q_{1k} = \mu_s \mu_z \omega_0 B = 0.8 \times 1.02 \times 0.35 \times 6 = 1.71 \text{kN/m}$$
$$q_{2k} = \mu_s \mu_z \omega_0 B = 0.5 \times 1.02 \times 0.35 \times 6 = 1.07 \text{kN/m}$$
$$q_1 = r_0 q_{1k} = 1.71 \times 1.4 = 2.39 \text{kN}$$
$$q_2 = r_0 q_{2k} = 1.07 \times 1.4 = 1.50 \text{kN}$$

(2) The height from eave to roof changes a little, so take the wind pressure variation coefficient at the eave to represent the value of the whole roof.
$$H = 10.8 + 2.1 = 12.9\text{m}$$
$$\mu_z = 1 + \frac{1.13 - 1.00}{15 - 10} \times (12.9 - 10) = 1.08$$

When the horizontal wind load of the roof slope, it is advisable to take the h_2 into the formula:
$$\overline{\omega_k} = [(0.85 + 0.5)h_1 + (0.5 - 0.6)h_2]\mu_z \omega_0 B = 5.92 \text{kN}$$
$$\overline{\omega} = \gamma \overline{\omega_k} = 1.4 \times 5.92 = 8.29 \text{kN}$$

3.3.3 Internal Force Analysis

There are two methods for analyzing the internal forces of a single-storey bent frame structure: considering the overall space effect of the plant and without considering the above effect. This section mainly discusses the internal force analysis of the bent frame without considering the overall space function of the plant.

The single-storey plant bent frame structure is generally statically indeterminate structure, and the number of statically indeterminate is the same as the span. There are many methods for calculating statically indeterminate structures, and the force method is generally used. For equal height bent frame, the shear distribution method is relatively simple.

In the calculation sketch of bent frame, if the elevation of every column is the same (Figure 3.20 a), or the top elevation of the column is different, the top of the column is connected with inclined beam (Figure 3.21b), the above two types of bent are called equal high bent frame. According to the assumption that the rigidity of a beam is infinite, the horizontal displacements of the column at both ends of the beam are equal when the bent frame is loaded. According to the characteristics of equal height bent frame, column shear force of each column can be worked out by shear distribution method, then the internal forces of the arbitrary cross section can be calculated under the condition of known shear force and external load of the independent cantilever column.

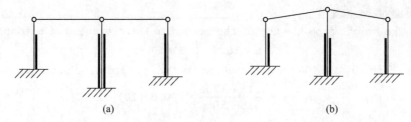

FIGURE 3.20 Equal height bent frame

3.3.3.1 Calculation of equal height bent frame by shear distribution method

In structural mechanics, when the top of single stage cantilever column has unit horizontal loads (Figure 3.21), the top horizontal displacement can be calculated as:

$$\delta = \frac{H_2^3}{3EI_2}\left[1+\lambda^3\left(\frac{1}{n}-1\right)\right] = \frac{H_2^3}{EI_2 C_0} \tag{3-11}$$

where, $\lambda = \dfrac{H_1}{H_2}$, $n = \dfrac{I_1}{I_2}$, $C_0 = \dfrac{3}{1+\lambda^3\left(\dfrac{1}{n}-1\right)}$, C_0 can be found according to Figure C.1 in Appendix C.

To get one unit of horizontal displacement at the top of column, the horizontal force $\dfrac{1}{\delta}$ needs to be applied (Figure 3.21). When the material is the same, the larger the cross section of the column is, the greater horizontal force requires. $\dfrac{1}{\delta}$ reflects the columns

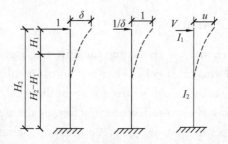

FIGURE 3.21 Shear stiffness of a single cantilever column

ability of resisting lateral displacement, which is often called the lateral stiffness of columns. The horizontal force V that produces the horizontal displacement u is column shear force, $V = u \frac{1}{\delta}$ (Figure 3.21).

(1) The horizontal concentrated force F acting at the top of the column

According to Figure 3.22, assuming the number of bent columns is n, the shear stiffness for each column i is $\frac{1}{\delta_i}$. The column shear force V_i can be calculated by the froce equilibrium condition and deformation compatibility condition.

$$V_i = \frac{1}{\delta_i} \times u, \sum_{i=1}^{n} V_i = \sum_{i=1}^{n} \frac{1}{\delta_i} \times u = F, u = \frac{1}{\sum_{i=1}^{n} \frac{1}{\delta_i}} F, V_i = \frac{\frac{1}{\delta_i}}{\sum_{i=1}^{n} \frac{1}{\delta_i}} \times F = \eta_i F$$

(3-12)

where, η_i—the shear distribution coefficient of column i. It is the ratio of the shear stiffness of itself to the sum of the shear stiffness of all columns.

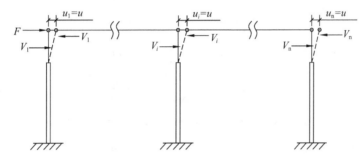

FIGURE 3.22 The shear distribution when horizontal concentrated load acting on the top of the column

After the shear distribution coefficient η_i is calculated, the shear force of each column can be calculated according to Formula (3-12). Then, according to the shear force of independent cantilever columns, the internal force can be calcualted out. The internal force calculation method of bent frames is called shear distribution method.

(2) arbitrary load acting on bent frame column

It can be seen from Figure 3.23 that the effect of any arbitrary load acting on the bent column is different from the horizontal concentrated force acting on the top of the column, the shear distribution method cannot be directly used in this case. In order to utilize the shear distribution coefficient, it can be calculated according to the following three steps:

① Add a fixed hinge support on the top of the bent column to prevent lateral displacement, and calculate its horizontal reaction force R.

② Withdraw the fixed hinge support and add the horizontal reaction force R with revelse direction on the top of the bent column to restore the original condition.

③ The superposition of the reaction force obtained in the above two steps is the the actual internal force of the bent frame.

The reaction force R of the fixed hinge support under various loads can be checked from Figure C. 2 to Figure C. 7 in Appendix C.

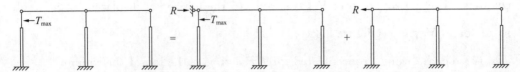

FIGURE 3.23 Shear distribution under arbitrary load

3.3.3.2 Calculation of internal forces of bent frames under common loads

The common load forms of a single-storey factory buildings are crane loads (includes vertical and horizontal loads) and wind loads. The internal forces of bent frames under these loads can be calculated according to the above-mentioned shear force distribution method.

(1) Calculation of internal forces of bent frame under vertical crane load

When the vertical loads of crane D_{max} and D_{min} acts on the corbels of the two columns of the bent frame at the same time shown in Figure 3.24, its internal force can be calculated out by the principle of force superposition.

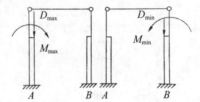

FIGURE 3.24 Calculation sketch under the action of D_{max} and D_{min}

When the vertical load of the crane D_{max} acts on the column A (Figure 3.24), only axial force produces on the lower pillar, the bending moment and shear force of the bent column section can be calculated according to the steps in Figure 3.25.

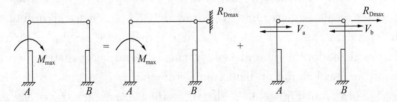

FIGURE 3.25 Internal force calculation of bent frame when M_{max} acts on column A

① Add a fixed hinge support on the top of the bent column, and calculate the counter force R_{Dmax} on the top of the column and the internal force of column under the M_{max} acted on the top face of the corbel. R_{Dmax} can be calculated by following formula.

$$R_{Dmax} = C_2 \frac{M_{max}}{H_2} \tag{3-13}$$

where, C_2—counter force coefficient of the top of the column, can be found from Figure C.3 in Appendix C.

② In order to balance the effect of additional fixed hinge support, the counter force on the top of the column R_{Dmax} is applied reversely at the top of the hinged bent column which produces lateral displacement. The Column shear force V_i can be calculated as cantilever column by following formula.

$$V_i = C_2 \frac{\frac{1}{\delta_i}}{\sum \frac{1}{\delta_i}} R_{Dmax} = \eta_i R_{Dmax} \qquad (3\text{-}14)$$

$$\delta_i = \frac{H_2^3}{C_0 E I_2} \qquad (3\text{-}15)$$

where, C_0—the column displacement coefficient under the action of unit horizontal load, can be found from Figure C. 1 in Appendix C.

③ Superimpose the internal forces of the columns obtained by the above two steps to achieve the internal forces of the bent columns.

Similarly, the internal force of the bent column D_{min} can be calculated when the vertical load of the crane is acting on column B. When the crane vertical load D_{min} acts on the column A and D_{max} acts on the column B, the internal force of the bent column can also be calculated by the above method. However, the internal force diagram of bent column is exactly opposite to that of column A, therefore, no need to be recalculated.

(2) Calculation of Internal Forces of Bent Frame under Horizontal Crane Load

The internal force of bent column under the action of crane horizontal load T_{max} can also be calculated by using the additional fixed hinged support and the shear force distribution method. The calculation diagram of the bearing action of the two-span equal height bent frame is shown in Figure 3. 26a. The internal forces of bent columns can be obtained by the superposition of the internal forces shown in Figure 3. 26b and Figure 3. 26c.

For the case shown in Figure 3. 26 b, the counter force on the top of the column R_i and the internal force of column can be calculated by the assumption that the single column with variable cross section is fixed hinged at the top and fixed at the bottom. The counter force on the top of the column R_i is obtained by:

$$R_i = C_3 T_{max}, \quad R_T = \sum R_i \qquad (3\text{-}16)$$

where, C_3—the column counter force coefficient under the action of crane horizontal load, can be found from Figure C. 4 to Figure C. 6 in Appendix C;

R_T—total counter force of bent frame column.

As shown in Figure 3. 26c, the column shear force V_i under the action of R_i can be calculated by shear distribution method. Then, internal force of column can also be calculated out.

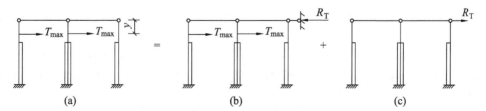

FIGURE 3. 26 Calculation of internal force of two span equal height bent under T_{max}

When the single span symmetric bent is used, the internal forces of each column under the action of T_{max} can be calculated out according to the cantilever column shown in

Figure 3.27.

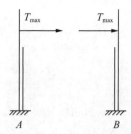

FIGURE 3.27 Internal force calculation of bent frame under T_{max}

Similarly, when the action of T_{max}'s direction is left, the internal forces of bent columns can also be calculated according to the above method.

(3) Calculation of internal forces of bent frame under wind load

Under action of the wind load F_w, q_1 and q_2 (Figure 3.28), the internal forces of the equal height bent frame can be obtained by the superposition of three forces in Figure 3.29a, b and c.

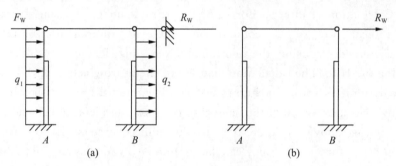

FIGURE 3.28 Effect of wind load on single span bent frame

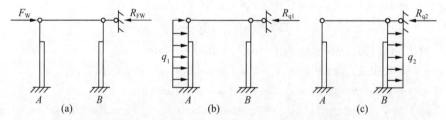

FIGURE 3.29 Internal force calculation of single span bent under wind load

① The calculation under the action of F_w: Figure 3.29a shows that the column counter force R_{FW} is identical to the F_w, and the column does not produce internal force.

② The calculation under the action of q_1: as shown in Figure 3.29b, the internal force of the column B is zero without no load. For column A, it is necessary to calculate the reaction force R_{q1} at the top column and its internal force of the column which upper end support is an immovable hinge, and the lower end is fixed. The reaction force of top column R_{q1} is:

$$R_{q1} = C_4 H_2 q_1 \tag{3-17}$$

where, C_4—the column counter force coefficient under the action of wind load, can be found from Figure C.7 in Appendix C.

③ The calculation under the action of q_2: For the case shown in Figure 3.29c, the analysis and calculation can be done in the same way as that of q_2. the reaction force of top column R_{q2} is:

$$R_{q2} = C_4 H_2 q_2 \tag{3-18}$$

Superimposing the internal force of bent column and column counter force under the action of F_W, q_1 and q_2, the internal forces of bent column and the total counter force on the top of the column R_W can be achieved (Figure 3.28a).

Under the reverse action of R_W as shown in Figure 3.28b, the internal force of the bent column can be calculated by the shear distribution method. Finally, superimposing the internal forces of the bent frame in Figure 3.28 a and b, we can get the internal force in the column of the original single span bent frame under wind load.

In conclusion, the above description is the method for calculating the internal force of the equal height bent without considering the overall space of the workshop. For unequal height and multi span bent frame, the internal force can be calculated by force method described in structural mechanics.

EXAMPLE 3.3

As shown in Figure 3.30, the left column A and right column B bear bending moment $M_1 = 160\text{kN} \cdot \text{m}$ and $M_2 = 141\text{kN} \cdot \text{m}$ respectively, and the moment inertia of cross section of the bent column are:

$I_1 = 2.13 \times 10^9 \text{mm}^4$, $I_2 = 14.38 \times 10^9 \text{mm}^4$, $I_3 = 7.20 \times 10^9 \text{mm}^4$, $I_4 = 19.50 \times 10^9 \text{mm}^4$.

All E_c are the same. Calculating the internal force of the bent column by shear force distribution method.

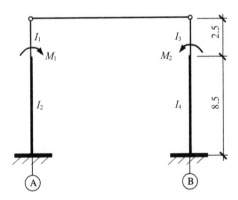

FIGURE 3.30 The diagram of the bent frame

SOLUTION

(1) Calculate n and λ:

Column A:

$$\lambda = \frac{H_1}{H_2} = \frac{2.5}{11} = 0.227, \quad n = \frac{I_1}{I_2} = \frac{2.13}{14.38} = 0.148$$

Column B:
$$\lambda = \frac{H_1}{H_2} = \frac{2.5}{11} = 0.227, \quad n = \frac{I_3}{I_4} = \frac{7.20}{19.50} = 0.369$$

(2) Calculate the column shear force with a fixed hinge support on the top of the column.

Add a fixed hinge support on the top of column A and B. Check the corresponding Table D-3 according to the frame column force.

Column A: $C_2 = 1.32$, Column B: $C_2 = 1.36$

Reaction force of fixed hinge support:

Column A:
$$R_A = \frac{M_1}{H_2} C_2 = \frac{-160}{11} \times 1.32 = -19.20 \text{kN}(\leftarrow)$$

Column B:
$$R_B = \frac{M_2}{H_2} C_2 = \frac{141}{11} \times 1.36 = 17.43 \text{kN}(\rightarrow)$$

So, the column shear force of column A and column B are:
$$V_{A.1} = R_A = -19.20 \text{kN}(\leftarrow)$$
$$V_{B.1} = R_B = 17.43 \text{kN}(\rightarrow)$$

(3) Shear distribution when the fixed hinge support is withdrawn

Add the horizontal concentrated force $-R_A$ and $-R_B$ on the top of the column. Then, calculate the shear force distribution coefficient of column A and column B according to C_0 found in Table D.1.

Column A: $C_0 = 2.80$, Column B: $C_0 = 2.95$

$$\delta_A = \frac{H_2^3}{C_0 EI_2} = \frac{11^3}{2.80 \times 14.38 \times E_c} = 33.06 \frac{1}{E_c} \text{mm}$$

$$\delta_B = \frac{H_2^3}{C_0 EI_4} = \frac{11^3}{2.95 \times 19.50 \times E_c} = 23.14 \frac{1}{E_c} \text{mm}$$

$$\eta_A = \frac{\frac{1}{33.06}}{\frac{1}{33.06} + \frac{1}{23.14}} = 0.412$$

$$\eta_B = 1 - \eta_A = 0.588$$

Under the action of $-R_A$ and $-R_B$, the shear force of each column is as follows:
$$V_{A.2} = 0.412 \times (19.20 - 1.43) = 0.729 \text{kN}(\rightarrow)$$
$$V_{B.2} = 0.588 \times (19.20 - 1.43) = 1.041 \text{kN}(\rightarrow)$$

Superimpose the state of (2) and (3).
$$V_A = V_{A.1} + V_{A.2} = -19.20 + 0.729 = -18.47 \text{kN}(\leftarrow)$$
$$V_B = V_{B.1} + V_{B.2} = 17.43 + 1.041 = 18.47 \text{kN}(\rightarrow)$$

(5) According to the bending moment and shear force on the top of the column, the internal force diagram can be calculated out and draw in Figure 3.31.

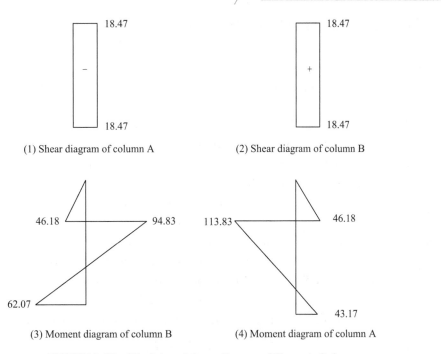

FIGURE 3.31　The internal force diagram of Example 3.3

3.3.4　Space Effect in Internal Force Analysis of Single-Storey Bent Frame Structure

3.3.4.1　The concept of space effect in a single storey factory building

The internal force analysis of the structure using plane bent frame instead of the entire frame structure is adapted to the situation shown in Figure 3.32a. In this situation,

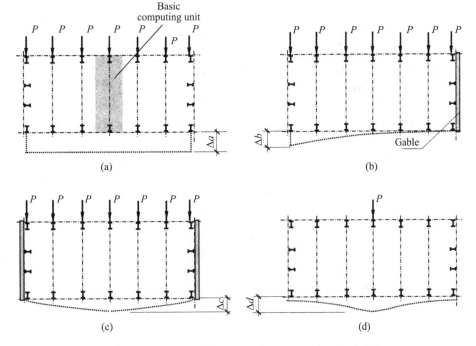

FIGURE 3.32　Space effect in a single-storey factory building

the displacement produced by the bent frames Δ_a are not equal, and there is no mutual restraint between the bent frames. However, for the three cases in Figure 3.32 b, c and d, the mutual restraint between the bent frame and the displacement of columns of each bent frame is different. At this point it is conservative to calculate the structure using the plane bent frame.

When the stiffness of each row of the single-story plant is different, or the load of each row frame is not the same, their displacement under the action of load will be restricted by other bent frames. The interaction between the bent frames is called space effect in a single-storey factory building.

3.3.4.2 Space effect distribution coefficient of single-storey plant structure

As shown in Figure 3.33, the bent structure under the action of horizontal load P produces the different horizontal displacement because of connection effect of longitudinal members. The horizontal force of the bent frame C is not P, but P_1. The other part of the load P_2 is taken up by other bent frames:

$$P = P_1 + P_2 \tag{3-19}$$

The space acting distribution coefficient of the bent frame is given by following formula:

$$m = \frac{P_1}{P_2} = \frac{\Delta_1}{\Delta} \tag{3-20}$$

where, Δ_1—the actual displacement of the bent frame C under horizontal load P;

Δ—the displacement of each row of the bent frame C under horizontal load P.

The spatial distribution coefficient m relates to some factors such as the roof stiffness, bent frame stiffness, plant span, plant length, the presence of gables in the temperature zone, and the tonnage and number of the crane. The specific values of m can be found in Table 3.2 and Table 3.3.

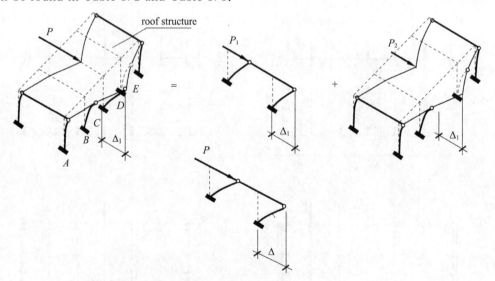

FIGURE 3.33 Displacement of single-storey factory building under horizontal load

TABLE 3.2 The spatial distribution coefficient of no purlin single span workshop

Factory condition	Tonnage of the crane t/kN	Plant length h/m			
		≤60		>60	
		span/m			
		12~27	>27	12~27	27
No gable at both ends or Gable at one end	≤75(735)	0.90	0.85	0.85	0.80
Gable at both ends	≤75(735)	0.80			

TABLE 3.3 The spatial distribution coefficient of single span workshop with purlins

Factory condition	Tonnage of the crane t/kN	Plant length/m	
		≤60	>60
No gable at both ends or Gable at one end	≤30(294)	0.90	0.85
Gable at both ends	≤30(294)	0.85	

The application areas of Table 3.2 and Table 3.3 are:

(1) When the roof structure is a large no purlin roof system with rib roofing panel ($h \geqslant 150$mm) and the connection between slab and roof truss is three-point welding, m can be set according to Table 3.2.

(2) When the roof structure is reinforced concrete structure or prestressed concrete tile, roofing tile, etc. and the connection between the purlin and the roof truss is welding, m can be adopt according to Table 3.3.

(3) The spatial calculation (i.e. $m=1$) is not considered in the calculation of the bent frame in the following cases when:

There is gable at one end of the plant or no gable at both ends and the length of the plant is less than 36m.

The skylight frame span is greater than one-half of the span of the plant or the layout of skylight frame makes the roof of the plant discontinuous in the longitudinal direction.

The column spacing is larger than 12m, or less than 12m, but columns spacing range are different and the maximum column spacing is greater than 12m.

The last quarter of the roof truss is flexible rod.

(4) The following requirements should be also considered when applying Table 3.2 and 3.3:

The gable should be solid brick wall. If there is a void on the gable, the weaken area of the horizontal section of the gable should not be greater than 50% of the total horizontal area of the gable. Otherwise, it should be regarded as no gable wall.

When the plant is equipped with expansion joints, the length of the plant should be taken as a unit of expansion joints. In this case, contraction joints should be considered as non-gable.

For the equal height multi-span plant, the spatial distribution coefficient should be calculated as follows:

$$\frac{1}{m} = \frac{1}{n_s}\left(\frac{1}{m_1'} + \frac{1}{m_2'} + \cdots + \frac{1}{m_n'}\right) = \frac{1}{n_s}\sum_{i=1}^{n}\frac{1}{m_i'} \qquad (3\text{-}21)$$

where, m—the spatial distribution coefficient of equal height multi span plant;

n_s—number of the bent span;

m_i'—the single-span space acting distribution coefficient of i span, can be found in Table 3.2 and 3.3;

m_i'—the single-span spatial distribution coefficient of i span, can be found in Table 3.2 and 3.3.

3.3.5 Internal Force Combination

3.3.5.1 Control section of column

Under the action of load, the internal force of the column varies along the height, several sections are selected to find the most unfavorable combination of internal forces, these sections are called controlling sections which determine the design of column and foundation. In general, for a single column, for the convenience of construction, the reinforcement of the upper column and lower column are different. Therefore, the control section of upper and lower column should be found respectively.

For the upper column, the bending moment and the axial force of the bottom section I - I are larger than the other sections, usually the column bottom is generally taken as the control section of the upper column (Figure 3.34). For the lower column, the bending moment usually gets the maximum at the corbel surface under the action of vertical load of the crane, but under the action of wind load and the horizontal load, the maximum value is located at the column bottom, so the corbel plane (II - II section) and the column bottom (III - III section) are generally taken as control sections of the lower column (Figure 3.34). When the column has a large concentrated load such as suspended wall gravity acting on the upper column, the cross section of the concentrated load can be selected as the control section according to its internal force.

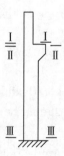

FIGURE 3.34　The control section of column

3.3.5.2 Effect of load combination

There is very low possibllity that several loads appear and reach the maximum at the same time. According to the stipulations of "*Load Code for the Design of Building Structures*" (GB 50009)(Ref. 3.4), when the structure of single-storey factory building is under the action of permanent load, roof variable load, wind load and crane load, the internal force calculation of each control section of the frame column can adopt the following simplified combination:

(1) Permanent load + any one live load

$$S = \gamma_G S_{GK} + \gamma_{Q1} S_{Q1K} \tag{3-22}$$

It includes the following three situations:

$$1.2 \times S_{GK} + 1.4 \times (S_{QLK} + S_{QHK})$$
$$1.2 \times S_{GK} + 1.4 \times S_{QWK}$$
$$1.2 \times S_{GK} + 1.4 \times S_{QHK}$$

(2) Permanent load $+0.9 \times$ (the sum of any two or more than two live loads)

$$S = \gamma_G S_{GK} + 0.9 \sum_{i=1}^{n} \gamma_{Qi} S_{QiK} \tag{3-23}$$

It includes the following three situations:

$$1.2 \times S_{GK} + 0.9 \times 1.4 \times (S_{QLK} + S_{QWK} + S_{QHK})$$
$$1.2 \times S_{GK} + 0.9 \times 1.4 \times (S_{QWK} + S_{QHK})$$
$$1.2 \times S_{GK} + 0.9 \times 1.4 \times (S_{QLK} + S_{QWK})$$

where, S_{GK} —load effect value calculated by standard value of permanent load;

S_{QLK} —load effect value calculated by standard value of variable roof load;

S_{QWK} —load effect value calculated by standard value of wind load;

S_{QHK} —load effect value calculated by standard value of crane load.

The most unfavorable value in the above combination is the internal force of the control section. The above simplified combination is of less theoretical basis and only suitable for manual calculation at the planning stage. All load effect combinations can be found in "*Load Code for the Design of Building Structures*" (GB 50009)(Ref. 3.4).

3.3.5.3 Cautions

(1) In any case, the internal forces generated by the permanent load must be considered.

(2) The vertical load of the crane D_{max} (or D_{min}) may act on the left column or right column of the same span of the plant, choose one case for combination.

(3) The horizontal load of the crane T_{max} may act on the left or right column of the same span of the bent frame, chose one case for combination.

(4) Since the horizontal load of crane cannot exit without vertical load, if the internal force generated by T_{max} is used, the corresponding internal force generated by D_{max} (or D_{min}) must be taken. The horizontal load does not necessarily occur when a crane has a vertical load in a span, so the internal forces generated by T_{max} must not be combined when the internal forces generated by D_{max} and D_{min}. Considering the characteristics that T_{max} direction may be left or right, the internal forces generated by D_{max} or D_{min} are usually combined with the internal forces generated by the corresponding T_{max} (only one item for multi span) to get the most unfavorable combination of internal force.

(5) The wind loads direction may be left or right, choose one case to participate in the combination.

3.3.5.4 The most unfavorable internal force combination

The bent columns of single-storey factory building are eccentrically compressed members, and generally adopts symmetric reinforcement. The shear force V has small impact on the designed reinforcement of the column. It is determined by bending moments M and axial forces N. Due to the large number of combinations of bending moments and axial forces, it is not easy to find out which combination of internal forces plays a controlling role and the combined results needs to be compared. For each control section of a rectangular, I-shaped column, generally the following four unfavorable combinations of adverse internal forces should be considered:

① Maximum positive bending moment M_{max}, corresponding axial force N and shear V.

② Maximum negative bending moment $-M_{max}$, corresponding axial force N and shear V.

③ Maximum axial force N_{max}, the corresponding bending moment M_{max} or $-M_{max}$ and shear V.

④ Minimum axial force N_{min}, the corresponding bending moment M_{max} or $-M_{max}$ and shear V.

For combination ① and ②, the corresponding axial force is determined when the moment is taken as the maximum positive or negative maximum. For the combination of ③ and ④, when the axial force is taken as the maximum value or the minimum value, the corresponding bending moment may be more than one. This is because when the wind load or the horizontal braking force of the crane acts, though the axial force may be zero, bending moment can also generate. Therefore, the corresponding maximum positive or negative bending moment must be taken.

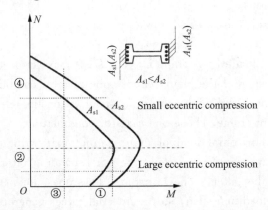

FIGURE 3.35   Relationship between bending moment, axial force and reinforcement ($A_{s1} < A_{s2}$)

3.4 Bent Frame Column of Single-Storey Factory

There are two kinds of columns in single-storey factory buildings: bent column and wind resistant column. The bent column can be designed according to the analysis of the internal force of bent frame.

3.4.1 Column Form

Reinforced concrete bent column generally consists of the upper column, the lower column and the corbel. Its structural forms can be classified into two types: single limb column and double limb column (in some cases, four limb columns can be used). The upper column is usually a solid rectangular section or a hollow ring section. There are many different sections of the lower column, as shown in Figure 3.36.

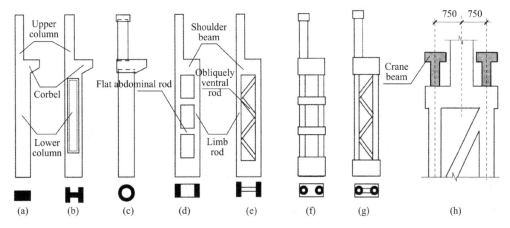

FIGURE 3.36 Structural forms of bent column

(1) Rectangular section column

Rectangular section column is generally used for single limb column (Figure 3.36a). The disadvantage of rectangular section is using much materials and has large self-weight. However, due to its simple structure and convenient construction, it is often used for small-scale industrial buildings. The section height of the rectangular cross-section column is usually ≤700mm.

(2) I-shape section column

The I-shaped section column are often used for single limb column (Figure 3.36b) of single storey factory buildings. The cross-sectional shape can make full use of the compressive capacity of concrete and resist the eccentric compressive load in the column. It has good overall performance and the construction is simple. The cross-section height of I-shape section column is generally from 800 to 1600mm.

(3) Double limb column

The lower column of a double limb column consists of a limb, a shoulder beam and a web member (Figure 3.36d,e). Due to the reasonable arrangement of members, the strength of concrete can be fully utilized. When the tonnage of the crane is large, the crane beam can be supported at the axis of the column limb to improve the force of the shoulder beam (Figure 3.36h). The cross-section height of double limb column is generally more than 1600mm. It is better to adopt the double limbs with diagonal web members, when the horizontal load is larger(Figure 3.35e).

(4) Tubular column

There are two kinds of tubular columns: circular tubular column and square tubular column. They can be made into single limb column, double limb column and four limb columns (Figure 3.36c, f, g). Among them, the application of double limb tubular column is more common. Tubes are typically molded on a centrifugal tube machine. Tubular columns have many advantages, such as high concrete quality of the column, high mechanization, and less lightweight. However, the current limitations of the centrifugal pipe machinery make it difficult to be widely used.

The column cross sectional size is determined by the lateral stiffness requirement of the workshop. Generally, if the size of rectangle or I-shape section can meet the requirements of Table 3.4, the lateral stiffness of the workshop can be guaranteed without checking the horizontal displacement. Otherwise, the horizontal displacement should be checked.

TABLE 3.4 Rectangular and I-shape section size selection reference table (column spacing is 6mm)

Types of columns	Section size			
	b	h		
		$Q \leqslant 10t(98kN)$	$10t(98kN) < Q \leqslant 30t(294kN)$	$30t(294kN) < Q \leqslant 50t(490kN)$
Lower column of plant with crane	$\geqslant H_1/25$	$\geqslant H_1/14$	$\geqslant H_1/12$	$\geqslant H_1/10$
Open-air crane column	$\geqslant H_1/25$	$\geqslant H_1/10$	$\geqslant H_1/8$	$\geqslant H_1/7$
Single-span plant without crane	$\geqslant H_1/30$	$\geqslant 1.5H/25$		
Multi-span plant without crane	$\geqslant H/30$	$\geqslant 1.25H/25$		
Gable wind column (Only bear the wind load and deadweight)	$\geqslant H_b/40$	$\geqslant H_1/25$		
Gable wind column (also bear the wall weight submitted by tie beam)	$\geqslant H_b/30$	$\geqslant H_1/25$		

H_1—the height from the top of the foundation to the bottom of fabricated crane beam or top column of place crane beam cast at site.

H—the total height of the column calculated from the top of the foundation.

H_b—the height from the top of the gable wind column foundation to the supporting point out of plane.

EXAMPLE 3.4

A single-storey industrial building uses bent frame structure, with 18m span and 6m column spacing. The plant is equipped with crane, the crane maximum wheel pressure standard value is $P_{max,k}$ = 110kN, the minimum standard wheel pressure value is $P_{min,k}$ = 30kN, the car wheel spacing is 4.5m. Draw the influence line diagram of bearing reaction of crane beam, and calculate the design value of the vertical load D_{max}, D_{max} of the designed bent frame column. (γ_Q = 1.4)

SOLUTION

(1) The influence line of bearing reaction of crane beam is shown in Figure 3.37.

$$y_1 = 1.0, y_2 = \frac{6-4.5}{6} = 0.25$$

(2) $D_{max} = \gamma_Q P_{max,k}(y_1 + y_2) = 1.4 \times 110 \times (1 + 0.25) = 192.5\text{kN}$

(3) $D_{min} = \gamma_Q P_{min,k}(y_1 + y_2) = 1.4 \times 30 \times (1 + 0.25) = 52.5\text{kN}$

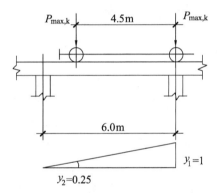

FIGURE 3.37 The influence line of bearing reaction of crane beam

EXAMPLE 3.5

The bent frame structure of a single-storey factory building bears wind load and the wind coefficient is shown in Figure 3.38. The basic wind pressure ω_0 = 0.4kN/m³, and the height of the column is + 11.8m. The Outdoor natural height is − 0.3m, and bent spacing B = 6 m. Calculate the wind load design value on the top of the bent frame. (μ_z = 1.0 at the height of 10m, μ_z = 1.14 at the height of 15m. μ_z at other height can be calculated by interpolation method.

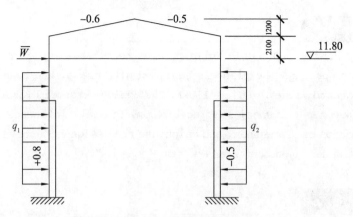

FIGURE 3.38　The diagram of Example 3.5

SOLUTION

The wind load above the top of column can be considered as the horizontal concentrated load applying on the top of the bent column. The wind pressure height variation coefficient can be calculated depending on the height from the outdoor terrace to the eaves: $11.8 + 0.3 + 2.1 = 14.2\text{m}$

$$\mu_z = 1.0 + \frac{1.14 - 1.0}{15 - 10} \times (14.2 - 10) = 1.118$$

$$\overline{W} = 1.4[(0.8 + 0.5)h_1 + [(0.5 - 0.6)h_2]\mu_z \omega_0 B = 9.8\text{kN}$$

The wind load above the top of column can be considered as uniform linear load. The wind pressure variation height coefficient can be taken according to the height value from the outdoor terrace to the top of the column:

$$11.8 + 0.3 = 12.1\text{m}$$

$$\mu_z = 1.0 + \frac{1.14 - 1.0}{15 - 10} \times (12.1 - 10) = 1.059$$

Distributed windward and leeward wind loads are:

$$q_1 = 1.4 \mu_{s1} \mu_z \omega_0 B = 2.846\text{kN/m}$$

$$q_2 = 1.4 \mu_{s2} \mu_z \omega_0 B = 1.779\text{kN/m}$$

3.4.2　Design of Bent Frame Column

The design of reinforced Concrete Columns of a single-storey plant structure includes:

(1) Select form of columns: During the structural design stage, a suitable column form is selected based on the specific conditions of regional materials and construction by technical comparison and economic analysis.

(2) Determine column size: The column total height and the height of its components are determined according to the structure of the plant, the rail top elevation give by the process design, crane tonnage, and building uniform modulus requirements. The cross-section size of the column is determined by the stiffness requirements of the bent frame, the supporting requirements of the roof truss and the crane beam on the

column.

(3) Determine the reinforcement of column: Based on the internal force calculation and the combination of internal forces of the bent frame, the reinforcing steel bars are calculated to ensure the strength and meet the requirements of structure design, the strength and crack resistance during the hoisting stage also need to be checked.

(4) Design column corbel.

(5) Design structure of connections: When the column is connected with the roof truss, crane beam or brace members, the embedded members of the column should to be designed.

The calculation and combination of the internal forces of sections (1), (2) and (3) above have been described in detail in the previous section. This section focuses on the calculation of the column length, hoisting stage check and the design of the column corbel.

3.4.2.1 Calculation of column length

The calculation length of the column l_0 depends on the support conditions and the height of the column. Based on the theoretical analysis and engineering experience, when calculating the bending moment increasing coefficient η_s of the eccentric compression members, the length of the column of single-storey factory may be calculated based on the values shown in Table 3.5.

TABLE 3.5 Calculation length of single storey industrial building column with rigid roof and trestle bridge column of open crane

Column		Bent direction	Vertical bent direction		Remarks
			Column bracing	No column bracing	
Plant without crane	Single-span	$1.5H$	$1.0H$	$1.2H$	H—column height, H_u—upper column height H_l—lower column height
	Multi-span	$1.25H$	$1.0H$	$1.2H$	
Plant with crane	Upper column	$2.0H_u$	$1.25H_u$	$1.5H_u$	
	Lower column	$1.0H_l$	$0.8H_l$	$1.0H_l$	
Trestle bridge column of open crane		$2.0H$	$1.0H$		

Note: In above table, H is the overall column height that is calculated from the top surface of the foundation. H_l is the column lower height which is calculated from the top surface of the foundation to the bottom of the assembled crane beam or the top surface of the concrete crane beam cast-in-place. H_u is the column upper height which is calculated from the bottom of assembled crane beam or the top surface of cast-in-place crane beam.

In above table, the column calculation length of the crane plant can be adopted with the situation that plant is without a crane when the crane load is not considered, but the upper column calculation length is still adopted with the situation that plant has a crane.

In above table, The column calculation length in bent frame direction is suitable for $\dfrac{H_u}{H_l}$ is not less than 0.3, otherwise, the calculation length should be $2.5H_u$.

3.4.2.2 Corbel design

In the plant structure, the corbel extending from the side of the column is used to support the roof truss, crane beam and tie beams and other components. In addition to bear large vertical loads, the corbel sometimes is subjected to seismic loads and horizontal loads due to wind forces. Therefore, the corbel is an important part of the column and must be particularly designed.

(1) Classification of corbels

According to the horizontal distance a from the acting point of vertical load to edge of corbel column root, the corbel can be divided into two kinds (Figure 3.39): long corbel($a > h_0$) and short corbel ($a \leqslant h_0$). Here, h_0 is the effective height of the cross section of the root of the corbel. The former can be designed according to the cantilever beam, and the latter is a variable section deep beam and its calculation method will be explained in following context.

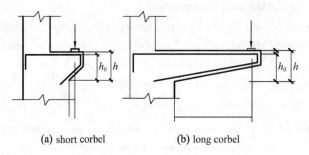

(a) short corbel (b) long corbel

FIGURE 3.39 Corbel categories

(2) Stress distribution and failure morphology of corbel

① Stress distribution

Through the photoelastic test of epoxy resin corbel model ($\frac{a}{h_0} = 0.5$), the main stress trace is shown in Figure 3.40. It can be seen from the diagram that the main tensile stress near the upper edge of the corbel is parallel to the upper edge, and the distance between the stress traces varies slightly, indicating that the tensile stress on the upper surface of the bracket is uniformly distributed along its length direction. The main compressive stress distribution near the bevel edge of the corbel is parallel to the line of ab. The spacing between the stress traces is also changed slightly, and the compressive stress distribution is even; in addition, there is a phenomenon of stress concentration near the junction of the upper column root and the corbel.

② Failure mode

The test of reinforced concrete corbel under vertical force shows that the crack development process and failure mode of corbel with different $\frac{a}{h_0}$ values are different:

Bending failure: When $1 > \frac{a}{h_0} > 0.75$, the longitudinal reinforcement ratio is low, the load reached the ultimate load of $20\% \sim 40\%$, the junction between the root of the upper column and the top of the corbel will appear top-down crack (1) (Figure 3.41a) firstly because of the stress concentration (Figure 3.40). These cracks are small and develops slowly, which has little effect on the mechanical properties of the corbel. When the load reaches the ultimate load of 40%~60%, the first diagonal crack (2) appears near the inner side of the loading plate, and its direction is basically along the main compressive stress trace. As the load continues to increase, the diagonal cracks continue to extend to the compression zone, the longitudinal stress increases continuously and gradually reaches the yield strength. At this time, the lateral diagonal cracks (2) rotate around the junction between the lower part of the corbel and the column, resulting in crushing of the crushed concrete in the compression zone. In the design, a sufficient number of longitudinal tensile reinforcement should be configured to avoid such failure.

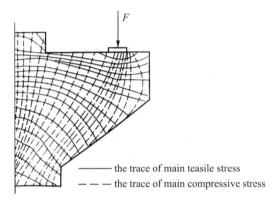

FIGURE 3.40　Stress state of corbel

Diagonal compression failure: When $0.75 \geqslant \dfrac{a}{h_0} \geqslant 0.1$, the initial fracture process is the same as the compression bending failure. As the loads continues to increase with the loads increasing, a lot of diagonal cracks (3) appear at the range of whole diagonal fracture zone at outside of diagonal cracks (2). When these cracks gradually penetrate, the concrete at diagonal pressure zone collapses, and the corbel fails (Figure 3.41b). Some corbels cracks do not appear (3), but a full-length crack (4) suddenly appears under the loading plate and the corbel is destroyed. These phenomena are called diagonal compression failure. When the failure occurs, the longitudinal tensile reinforcement reaches the yield strength. The bearing capacity calculation of corbel is mainly based on this failure mode.

Shear failure: When $\dfrac{a}{h_0} < 0.1$, though with a larger $\dfrac{a}{h_0}$ or the outer edge height of corbel is small, a series of short and thin oblique cracks appear at the joint of the lower column (Figure 3.41d), and finally the corbel is broken down from the column when they come into penetrating cracks. At this moment, the stress of the longitudinal tensile steel reinforcement in the corbel is small, this can be prevented by controlling the cross-section height of the corbel h and taking necessary construction measures.

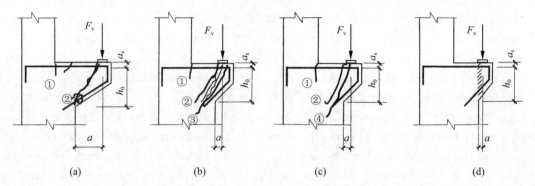

FIGURE 3.41 Failure morphology of corbel

These are the three main failure forms of the corbel. In addition, there are some failure phenomena such as the local compression failure of concrete under the loading plate and due to the small size of loading plate, and there may be pullout failure due to the bad anchorage of longitudinal tensile steel bars.

(3) Determination of dimension of corbel Section

The width of the corbel section is the same as that of the column, and the height of the corbel is determined to ensure that oblique cracks (2) do not appear in the normal service stage of the corbel or only a small number of fine cracks occur. According to the test results, the size of corbel should comply with the requirements of the following formula:

$$F_{vk} \leqslant \beta \left(1 - 0.5 \frac{F_{hk}}{F_{vk}}\right) \frac{F_{vk} b h_0}{0.5 + \dfrac{a}{h_0}} \tag{3-24}$$

where, F_{vk}, F_{hk}—the vertical force and the horizontal tensile force that calculated respectively according to the load standard combination at the top of the corbel;

β—the control factor of the crack: it takes the value 0.65 for the corbels of the crane beam and 0.8 for the other corbels;

a—the horizontal distance between the action point of vertical force and the edge of lower column. The install deviation 20mm should be considered at this time; when the vertical force acting line is still within the lower column section after considering the deviation of the 20mm installation, $a=0$ should be taken.

b—the width of corbels.

h_0—effective height of vertical cross section between corbel and lower column, $h_0 = h_1 - a_s + c_1 \tan\alpha$, when $\alpha > 45°$, the $\alpha = 45°$ can be taken.

The meaning of the rest of the symbols are shown in Figure 3.42.

When the $\dfrac{a}{h_0}$ is large, in order to prevent the diagonal cracks extend to the bottom diagonal plane of the corbel, and induce similar shear failure along vertical cross-section. The outer edge height h_1 of the corbel should not be less than $\dfrac{h}{3}$, and 200mm.

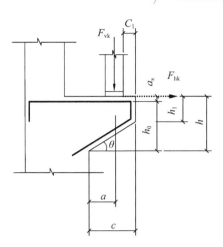

FIGURE 3.42 Corbel dimensions

In order to prevent the peeling of the protective layer, the distance c_2 between the outer edge of the leg and the outer edge of the crane beam should not be less than 70mm.

In order to prevent extreme stress concentration occur at the junction of the bottom of the corbel and the lower column with the appearing of the diagonal cracks. The bottom inclination should not be less than 45 degrees.

$$\sigma_c = \frac{F_{vk}}{A} \leqslant 0.75 f_c \qquad (3\text{-}25)$$

where, A—local compression area;

f_c—the design value of concrete axial compression strength.

When Formula (3-25) is not satisfied, effective measures should be taken to increase the compression area, improve the strength grade of concrete or set up the steel met.

(4) Reinforcement calculation and construction requirements of Corbel

The corbel can be regarded as a triangular truss with a top tensile steel bar as a horizontal pull rod (pulling force is $f_y A_s$) and a concrete inclined pressure belt as a compression bar, as shown in Figure 3.43.

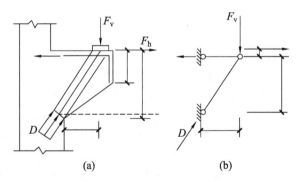

FIGURE 3.43 Calculation diagram of corbel

① Longitudinal reinforcement calculation and construction requirements of corbel

According to the calculation diagram shown in Figure 3.43, when the design value of vertical force and the horizontal pull force are acting on the corbel, the following

design expressions can be obtained considering the moment equilibrium of point A:
$$F_v a + F_h(\gamma_s h_0 + a_s) \leqslant f_y A_s \gamma_s h_0$$

Here $\gamma_s = 0.85$, $(\gamma_s h_0 + a_s)/\gamma_s h_0 = 1.2$, the area of the total cross section of the tensile steel bars A_s is obtained:

$$A_s = \frac{F_v a}{0.85 f_y h_0} + 1.2 \frac{F_h}{f_y} \tag{3-26}$$

where, a—horizontal distance from the point of action of the vertical force to the edge of the lower column, the installation deviation 20mm should be taken into account when $a < 0.3h_0$, and the value of a is $0.3h_0$;

h_0—the effective height of the cross section of the root of the corbel;

f_y—design value of longitudinal reinforcement.

HRB400 or HRB500 steel should be used for longitudinal tensile steel. The reinforcement ratio of longitudinal tensile reinforcement required for vertical force should be less than 0.2% and $0.45 f_t/f_c$ according to the effective section of corbel, and should not be greater than 0.6%. The number of roots should not be less than four, and the diameter should not be less than 12mm.

② Structural requirements for horizontal stirrup and flexural bar

The oblique section shear strength of corbel mainly depends on the concrete. The horizontal stirrups have almost no direct effect on the shear strength of the oblique section, but can inhibit the development of diagonal cracks and indirectly increase the shear strength of the oblique section. Bend steel bars also play an important role in suppressing the development of oblique cross-section cracks. Therefore, if the cross-sectional dimensions of the corbel satisfy the cracking conditions of Formula (3-24) and the horizontal stirrups and the flexural reinforcement bars are well distributed based on construction requirement. It is not necessary to calculate the bearing shear capacity of the diagonal section.

The diameter of horizontal stirrups should be 6~12mm and the spacing should be 100~150mm, and the total section area of horizontal stirrups in the range of $\frac{2}{3}h_0$ should not be less than the 1/2 of the total section area of tensile reinforcement under vertical load.

When the shear span ratio of the corbel is greater than 0.3, the bending bars should be set up. Bending bars can be HRB400, HRB500, HRBF400, or HRBF500 grade, and should be distributed in the range between $l/6$ to $l/2$ in the upper part of the corbel (Figure 3.44). The cross-sectional area should not be less than half of the tensile steel under the vertical force, the number should not be less than 2, and the diameter should not less than 12mm.

All longitudinal reinforced bar and bending reinforcement should be extended along the outer edge of the corbel into the column 150mm long and then truncated (Figure 3.43). The anchorage length of the longitudinal reinforced bar and the bent bar into the upper column should not be less than the anchorage length of the reinforced bar l_a. When the upper column size is insufficient, it should extend to the outside of the column and bend down. The horizontal projection length should not be less than $0.4l_a$. The vertical projection length should be $15d$. The anchorage length should be calculated from the edge of the upper column.

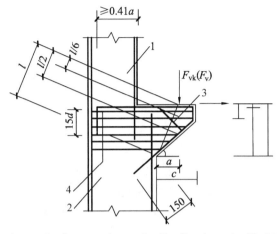

1—Upper column; 2—Lower column; 3—Bending bar; 4—Horizontal stirrup

FIGURE 3.44 Reinforcement structure of corbel

EXAMPLE 3.6

The width of reinforced concrete corbel is 400mm, the concrete grade is c30 (tensile strength standard value is 2.01N/mm²), the vertical load standard value at the top of the corbel is 150kN, the horizontal load standard value is 70kN and the crack control coefficient is 0.65 (shown in Figure 3.45). Check whether the cross section of the corbel satisfies the control conditions of the diagonal cracks.

$$a_s = 40\text{m}, \quad F_{vk} \leqslant \beta\left(1 - 0.5\frac{F_{hk}}{F_{vk}}\right)\frac{f_{tk}bh_0}{0.5 + \dfrac{a}{h_0}}$$

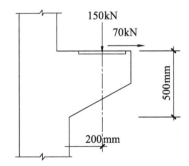

FIGURE 3.45 The size of corbel in Example 3.6

SOLUTION

$\beta = 0.65, b = 400\text{mm}, F_{hk} = 70 \times 10^3\text{N}, F_{vk} = 150 \times 10^3\text{N}$
$f_{tk} = 2.01\text{N/mm}^2, a = 220\text{mm}, h_0 = 500 - 40 = 460\text{mm}$

$$\beta\left(1 - 0.5\frac{F_{hk}}{F_{vk}}\right)\frac{f_{tk}bh_0}{0.5 + \dfrac{a}{h_0}} = 188.4 \times 10^3 \text{ N} > F_{vk}$$

The cross section of the corbel satisfies the control conditions of the slanting cracks

EXAMPLE 3.7

A corbel of the column is shown in Figure 3.46. The vertical load standard value is 300kN, the horizontal load standard value is 60kN and HRB335 steel bar is used ($f_y = 300\text{N/mm}^2$). Calculate the area of the longitudinal reinforced bar of the corbel.

($a_s = 60\text{mm}, A_s = \dfrac{F_v a}{0.85 f_y h_0} + 1.2\dfrac{f_h}{f_y}$, when $a < 0.3 h_0$, $a = 0.3 h_0$ can be taken)

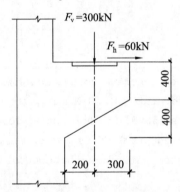

FIGURE 3.46 The size of corbel in Example 3.7

SOLUTION

$h_0 = 800 - 60 = 740\text{mm}$
$a = 200\text{mm} < 0.3 h_0 = 222\text{mm}$
So: $a = 0.3 h_0 = 222\text{mm}$

$A_s = \dfrac{F_v a}{0.85 f_y h_0} + 1.2\dfrac{f_h}{f_y} \approx 593\text{mm}$

3.4.2.3 Hoist inspection

The stress state of column during the hoisting process is different from the service stage, and the strength of the concrete may not reach the design strength at this time. Therefore, the bearing capacity and crack width of the column should be checked according to the force characteristics of the column during the hoisting process and the actual strength of the material.

There are two types of hoisting methods: turning hoisting and flat hoisting. Flat lifting is more convenient, and the turning hoisting means structural components like columns must be turned over before hoisting.

The calculation diagram of the column in the checking calculation should be determined according to the setting of the hoisting point. If using one-point lifting, lifting point is located at the corbel root, the most unfavorable stress stage in the hoisting process is the status when the hoisting point just leaves the ground. In this case, the bottom of the column is on the ground, and the column is a flexural member under the action of its self-weight.

When flat lifting is applied, the I-section can be simplified to a rectangular section, only the four flanges reinforcement of the two wings of are considered to participate in the work. If the longitudinal reinforcement is in the flange, it can also be considered. When the flat lifting does not meet the requirements of the bearing capacity or the limit value of the crack width, it can be used as turning hosting. Then the force direction and the application stage of the column section are consistent. The general bearing capacity and the crack width can meet the requirements without checking computation.

When checking the flexural capacity of the hoisting stage, the coefficient of the weight of the gravity load of the column is 1.2. Considering the dynamic effect of lifting, it should also be multiplied by a dynamic coefficient of 1.5. Because the time of the hoisting stage is shorter than the service stage, the structural importance coefficient is reduced to the lower level (usually 0.9). The actual strength of the concrete strength required when hoisting is usually more than 70% of the design strength.

When the carrying capacity or crack width requirements is not satisfied, the method of adjusting or adding the lifting point should be first adopted to reduce the bending moment or taking temporary reinforcement measures to solve it. When it is difficult to adopt these methods or measures, we can increase the strength grade of concrete or the number of longitudinal reinforcement.

3.5 Design of Crane Beam

Crane beam is an important component for single storey factory building. It directly bears the crane load and passes it to column of the bent frame, thus it plays an important role in the normal operation of the crane and the longitudinal stiffness of the workshop.

3.5.1 Mechanical Characteristics of Crane Beam

Crane beam is simply supported by columns. Its internal force characteristics depend on the loads' characteristics on the crane as the following:

(1) Crane load is movable load

The crane loads are two concentrated moving loads, moving vertical load P and horizontal load of Z. Therefore, not only the vertical bending under the action of self-weight and P, but also the two-way bending under the combined action of self-weight,

P and Z should be considered. For calculation of moving load, the maximum internal force of each section can be calculated by using the method of influence line or envelope diagram. When calculating the vertical load on crane beam, column corbel, the maximum or minimum wheel pressure should be used separately in addition to the unfavorable position of one and more cranes.

(2) Crane load is a cyclic load

During the service life of crane beam, the crane load maybe repeat several millions of times, thus the material strength of the structure or components will degrade when directly bearing the crane loads. Therefore, for the crane beam, besides static calculation, the fatigue checks must be carried out. When performing the fatigue check, the standard value and the dynamic coefficient should be considered for the crane load.

(3) Dynamic characteristics of crane load

The bridge crane of heavy working level with high speed has obviously dynamic effects on the crane beam. Therefore, when calculating the strength of the crane beam, its connection or the crack resistance of the crane beam, the vertical load of the crane should be multiplied by the dynamic coefficient. For suspension cranes and soft hook cranes with working level A1 to A5, the dynamic coefficient is 1.05. For the soft hook cranes, hard hook cranes and other special cranes with working level A6 to A8, the coefficient is 1.1.

(4) Consideration of the eccentricity of the crane load

The vertical load P and the horizontal load Z of the crane have eccentricity effect on the bending center of the cross section of the crane beam. The eccentricity of the vertical load P is caused by the allowable error of the crane track installation. Under the two eccentric loads, the crane beam will be in the torsional state. Therefore, the crane beam is a bending, shearing and twisting component under two-way bending moment.

3.5.2 Selection of Crane Beam

At present, according to the material difference the commonly used crane beams in industrial buildings can be divided into three types: reinforced concrete, prestressed concrete and steel—concrete composite structure. It can further be divided into five types according to the beam shape: uniform T shape and I-shaped cross section crane beam, fish belly crane beam, folding crane beam, arch crane beam and arch truss crane beam.

The crane beam should be selected flexibly according to the span, weight, work system, material supply, technical condition and time limit of the crane. The following suggestions can be referenced according to the practical experience of the project:

(1) For 4m to 6m span crane beam, its lift weight less than 300kN for light and intermediate working level or 200kN for heavy working level, the reinforced concrete crane beam and prestressed concrete crane beam can both be used. The weight of the light and intermediate working system is more than 300kN, and the weight of the heavy grade work is more than 200kN, then the prestressed concrete crane beam should be used.

(2) For 9m span crane beam, the lifting weight is below 100kN and 100kN, both ordinary reinforced concrete crane beam and prestressed concrete crane beam can be used. If the lift weight of the intermediate or heavy working grade is more than 100kN, and prestressed concrete crane beam or truss crane beam shall be used.

(3) For the 12m and 18m span crane beams, the prestressed concrete crane beam and the truss type crane beam are generally used.

3.5.3 Internal Force Calculation of Crane Beam

The maximum value of bending moment and shear force of the crane beam are generally obtained by the method of influence line. As shown in Figure 3.45a, for the I—I section, according to the moment influence line of the section, the maximum moment (Figure 3.45b) can be obtained by considering four most unfavorable positions (Figure 3.46c~f). Connecting all maximum moment respectively by the line, the bending moment envelope of a crane beam can be obtained. It must be pointed out that the absolute maximum bending moment of the beam is not at the mid span of the beam. In the same way, the shear envelope of the crane beam can be obtained.

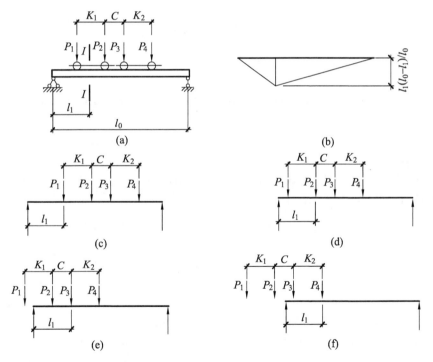

FIGURE 3.47 The internal force of a crane beam by means of the influence line method

The lateral horizontal dynamic force of the crane beam is on the top of the crane rail, with eccentricity e_z from the bending center of the crane. The installation error of the crane beam makes the $\mu_d P_{max}$ also has an eccentricity e_1 (generally $e_1 \leqslant 20mm$).

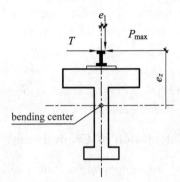

FIGURE 3.48 The load on the crane beam

The torque in the crane beam can be obtained (Figure 3.47):
$$T_t = \mu_d P_{max} e_1 + T e_z \tag{3-27}$$
The absolute maximum torque in a crane beam is also obtained by influence line method, which occurs near the support.

References

3.1 Rapid introduction to architectural design[M]. Beijing: China Electric Power Press, 2007.

3.2 Optimization and example of architectural structure design [M]. Beijing: China Construction Industry Press, 2012.

3.3 Dealing with the hot issues of architectural design [M]. Beijing: China Construction Industry Press, 2015.

3.4 Load code for the design of building structures[S]. GB 50009—2012. Beijing: China Construction Industry Press, 2012.

Questions

3.1 What are the common structures of a single-storey factory building?

3.2 What is the function of the column bracing?

3.3 What are the main loads of the single-storey factory structure?

3.4 What are the destructive forms of concrete corbel?

3.5 A single-storey factory building has a span of 30m, and the span of bent column is 6m. The basic wind pressure ω_0 is 0.42kN/m², the surface roughness category is B, the height from the top of column to the outdoor ground is 16m. Please calculate the wind load q_1, q_2 on the bent frame and their design value($\mu_z = 1.15$).

3.6 The corbel size and its design load are shown in the picture (unit: mm). The design value of the vertical force acting on the top of the corbel F_v is 640kN, the design value of horizontal tension F_h is 100kN. The steel bar is HRB400 ($f_y = 360$N/mm²).

$a_s = 50\text{mm}^2$. Try to calculate the area of longitudinal tensile steel bar A_s required on the top of the corbel.

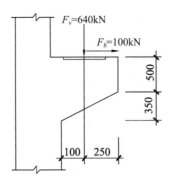

FIGURE 3.49　Sketch map of Example 3.6

Chapter 4

Multilayer Frame Structures

4.1 Frame Construction Types and Layout

4.1.1 Features and Types of Frame Structures

The characteristic of a frame structure is the conection of beam and column is rigid. It has better lateral stiffness than beat frame structure.

According to the difference of the construction method, the frame can be divided into cast-in-place frame, fabricated frame, and precast-monolithic frame.

The cast-in-place beam and column has a good integrity and seismic performance. The disadvantage is that it has the huge site operation workload, long construction period and large formwork demand. The assembled frame is a framework whose beam, column and floor are prefabricated and welded together to form a whole frame. Because all the components are prefabricated, standardization, industrialization and mechanization can be realized. The constrcuction of fabricated frame is quick during site construction, but it needs a great deal of transportation and hoisting work. This kind of structure has poor integrity and is weak in earthquake resistance, so it is not suitable to be applied in earthquake areas. The precast-monolithic frame means that the beams, columns and floors are prefabricated. After lifting and placing, the steel reinforcement is welded or tied, and the joint of the frame is formed by pouring the concrete at the site, and the components are connected into a whole. This kind of frame has good seismic capability and prefabricated components, and has the advantages of cast-in-place frame and fabricated frame. The disadvantage is that the construction on site is complicated, in addition, it is contains complex intermediate state.

4.1.2 Structure Layouts

4.1.2.1 Column arrangements

Column layout should not only meets the functional requirements of the construction, but also makes the structural stress reasonable and the construction convenient. If column arrangement is uniform and symmetrical, to structural stress, this can make the internal forces under vertical load uniform and fully make use of material strength, so it is beneficial to structure. The column arrangements should also consider modular and standardization component sizes and use less categories in order to meet the requirement of industrial production: efficiency and convenient construction. The fabricated structure should also consider the maximum length and maximum weight of the component to meet the restrictions of lifting and transportation.

4.1.2.2 Load bearing frame arrangements

For a typical building, vertical load is the main load, so how to resist the vertical load is the key point of structural layout. According to the transmission route of gravity load, the frame structure has three kinds of layout schemes: transverse load-bearing bearing frame (Figure 4.1a), longitudinal load-bearing frame (Figure 4.1b) and transverse-longitudinal horizontal load-bearing frame.

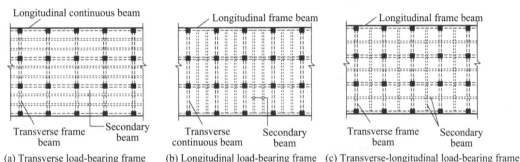

FIGURE 4.1 Frame load-bearing scheme

For transverse load-bearing scheme, the floor load is first transmitted to the transverse frame beam and then frame columns. Because the longitudinal beam does not bear the vertical load, the section of longitudinal tie beams can be smaller. Compared with the longitudinal frame beam, the transverse frame beam often has less span and poor lateral stiffness. The lateral bearing capacity of the building can be improved by adopting the transverse frame bearing scheme. Smaller longitudinal tie beams are conducive to lighting and ventilation of the house.

For longitudinal load-bearing scheme, the frame main beams are arranged longitudinally and the connecting beams are arranged transversely. Because the floor load is transmitted from the longitudinal beam to columns, the transverse beam is smaller in height, which facilitates the passage of the equipment pipeline; when the width direction needs a larger space in the house, higher ceiling height can be obtained. The disadvantage of the longitudinal frame bearing scheme is that the lateral stiffness of the building is poor.

For transverse-longitudinal bearing scheme, the main frame are arranyed in two directions to resist the floor load. This can make the building have large lateral rigidity in two directions, and the structural integrity performance is good. If the reinforced concrete multi-storey frame structure is built in a seismic fortification area, the combined transverse and longitudinal frame load scheme should be adopted no matter how the vertical load is transmitted.

4.1.2.3 Vertical arrangements

Vertical arrangements refers to the determination of the height of the structure and the change of the structure along the vertical direction. The functional requirements of the building are satisfied, and the vertical structure should be regular and simple. The common structural changes along the vertical direction are: basically unchanged along the vertical direction, which is a common and reasonable form of force, the ground floor with large space, such as a shopping center, the top floor with large space, such as the top floor for sightseeing room, conference room, dining place, and so on.

For a vertical regular structure, the layout of the structure is mainly the plane arrangements. When the vertical structure is very irregular, the arrangements of each different plane should be carried out separately. At this point, both the plane and the vertical layout have to be considered together.

In order to facilitate the structural force in the plane, the frame beam should be pulled through and straight on; in the vertical direction, the frame column should be up and down, the beam and column axis should be in the same vertical plane.

4.1.3 Structure Form Selection and Section Dimension Estimating

Structural selection includes determining the form and size of the components. Frames are generally higher order statically indeterminate structures, so it is necessary to determine the form and section size of the components before analysis.

For the selection of roof structure, refer to the comesponding content of Chapter 2 structure. The cross-section of the frame beam is usually rectangular. When the floor is cast-in-place, part of the floor can act as the flange of the beam, and the cross-section of the beam becomes T-shaped or L-shaped. When using precast floor slab, the cross section of the beam is usually cross shaped (Figure 4.2a) or a flower basket shape (Figure 4.2b) in order to reduce the height of the building and increase the headroom height; Laminated beams can also be used as shown in Figure 4.2c. The pre-fabricated beam is T-shaped section, after installation of the precast beam, the panel are installed in place, part of the concrete is poured in place to make the post-poured concrete and the pre-fabricated beamwork as a whole.

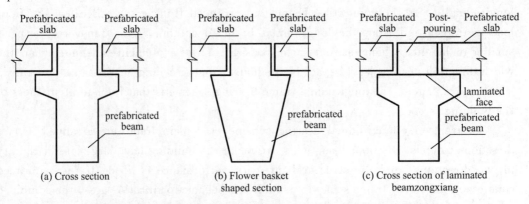

FIGURE 4.2 Section forms of prefabricated beams and laminated beams

The cross-section of a frame column is often rectangular or square. Sometimes due to architectural requirements, they can also be designed into circular, octagonal, T-shaped, and so on.

In contrast to the main beam in Chapter 2, the frame beam are designed to resist horizontal loads in addition to the vertical loads. Therefore, to determine the cross-sectional height of the frame beam, the horizontal load must also be considered besides the span, vertical load size, material strength and other factors. Under normal circumstances, the cross-sectional height can take $1/12 \sim 1/8$ of span height; section width can take the $1/3 \sim 1/2$ of cross-sectional height, and should not be less than 200mm. The cross-sectional dimensions of the frame beam can also be estimated from the vertical load. First, we take $M_{max}=(0.6 \sim 0.8)M_0$, $V_{max}=V_0$, M_0, V_0 are the design values of midspan bending moment and shear strength of a simply supported beams with the same vertical load respectively. Then, the following conditions must be satisfied:

$$M_{max} \leqslant \alpha_s f_c b h_0^2 \tag{4-1}$$

$$V_{max} \leqslant 0.235 f_c b h_0 \tag{4-2}$$

In the formula, α_s is 0.22 for the first seismic grade; for second and third-seismic grade, 0.29 is used.

The section sizes of concrete rectangular columns can approximately take the $1/18 \sim 1/12$ of height, $h_c=(1-2)b_c$ respectively, and can also be calculated by the following steps: estimate the axial force N_c of a column under vertical load according to the load area, (Approximately, load basic combination value is $12kN/m^2 \sim 18kN/m^2$), then take $N=(1.2 \sim 1.4)N_c$. Further, calculate the cross-sectional area according to $N=Af_c+\rho A f_y$ (ρ is the reinforcement ratio, 1% is desirable).

For seismic prone areas, the column cross-sectional area is generally controlled by the limit value of axial pressure ratio. For frames with different seismic grades, axial compression ratio should meet the following conditions:

$$\frac{N}{f_c A} \leqslant 0.65 \text{ (first seismic grade)}$$

$$\frac{N}{f_c A} \leqslant 0.75 \text{ (second seismic grade)}$$

$$\frac{N}{f_c A} \leqslant 0.80 \text{ (third seismic grade)}$$

$$\frac{N}{f_c A} \leqslant 0.90 \text{ (fourth seismic grade)} \tag{4-3}$$

The width and height of the column cross-section should not be less than 300mm, the ratio of the height to the width should not be greater than 3; the diameter of circular columns should not be less than 350mm.

4.2 Internal Force and Displacement Calculation

Multi-layer frame are generally statically indeterminate structures, and it is very complicated to analyze them by force method or displacement method of structural mechanics. This section introduces the simplified analysis methods commonly used in engineering, including the stratification method under vertical loads, the inflection point method under horizontal load and the D value method under horizontal load.

4.2.1 Analysis models

4.2.1.1 Calculation units

For two-direction orthogonal multi-storey frame structure, the horizontal loads can be divided into longitudinal horizontal load and lateral horizontal load respectively for analysis. When the structure and the load is uniform, the structure can be analyzed as the plane structure without considering the spatial action. For the regular framework shown in Figure 4.3, it is approximately assumed that the structure and the load are uniform in the longitudinal and transverse directions. Therefore, the shaded area can be taken as the unit of computation, i.e. half of both sides of the span.

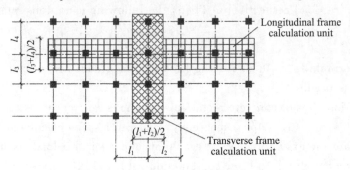

FIGURE 4.3　Frame calculation unit

4.2.1.2 Calculation diagrams

The calculation sketch includes the structural forms, axis dimensions, and cross-sectional characteristics.

(1) The structural form

The multi-layer frame column and the foundation generally adopts a rigid connection joint, that is, the column is fixed to the top surface of the foundation; and the nodes of the beam and the column are taken as rigid joints or hinged connection according to the structural arrangement.

(2) Axis size

In the structural calculation diagram, the bar is shown with its axis. The distance between the axis of the two adjacent columns determines the span of the frame l_i, and the distance between the upper and lower beam axis determines the height of the frame h_i. When the cross-sectional dimension of the column changes with the height, the axis of the column generally coincides with the centroid of the smaller cross-section; the axis of the beam coincides with the cross-sectional centerline of the beam, and when the height of each floor is the same, and the cross-sectional height of the beam is the same, the frame height is equal to the height of the floor.

(3) Cross-section characteristics

The bending stiffness of the section of the frame column can be calculated directly according to the shape and size of the section. When calculating the flexural rigidity of the frame beam, the influence of the floor slabs should be taken into account. When the floor is poured together with the frame beam, the floor forms the flange of the beam. For cast-in-place floor beams, the center frame is $I=2I_0$ and the side frame is $I=1.5I_0$. For the beams of assembled monolithic floor, the middle frame is taken as $I=1.5I_0$, the side frame is $I=1.2I_0$. For the beam of the assembled floor, $I=1.5I_0$. Here, I is the moment of inertia of the beam without considering the influence of the floor.

The schematic diagram of the frame structure is shown in Figure 4.4.

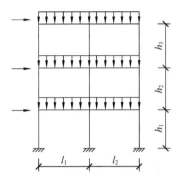

FIGURE 4.4 Frame calculation diagram

4.2.1.3 Loads calculation

Horizontal loads, including wind loads and horizontal seismic action, are generally simplified into horizontal concentrated forces acting on the joints.

Vertical loads include permanent loads and variable loads. The self-weight of beams, slabs, columns, walls and fixtures belong to permanent load, and the floor uniformly distributed loads, roof snow loads and floor area ash loads are variable loads.

The calculation method of the load transmitted from the roof and floor including the self-weight and distributed load on the lab is the same as that of the main beam in Chapter 2. When there are secondary beams, the frame beams are subjected to a concentrated load. If no secondary beam, the supporting beam receives uniform load in the case of one-way slab. For two-way slab, and the beam in short span direction bears a triangle-shaped distribution load on the frame beam and the long span direction bears a trapezoidal distributed load.

The self-weight of the infill wall and the frame beam directly acts on the beam in the form of a linear distribution load. The frame column takes its own weight and half of the height of the lower floor acts as a concentrated load on the beam and column joints.

4.2.2 Internal Force Analysis under Vertical Loads

Under the influence of vertical loads, the lateral displacement of the regular multi-story and multi-span frame structure is very small, and the lateral displacement is approximated zero. At this time, it is convenient to calculate and analyze the internal force using the moment distribution method or iterative methods. In the preliminary design, a more simplified hierarchical method can be used.

According to the structural mechanics, when the distal end condition is fixed, the bending moment transfer coefficient of a prismatic member is equal to 0.5; when the distal end is hinged, the moment transfer coefficient is 0. The transfer coefficient of the actual situation is between $0 \sim 0.5$. In the frame structure, the distal end of the component is usually connected to several members. The bending moment is distributed to several adjacent members, then transferred to the distal end of these members. The moment of the second pass is even smaller. Therefore, the hierarchical method assumes that the vertical load in a layer of frame beams only generates bending moment and shear force on the frame beams and frame column connected with these beams. The frame columns do not generate bending and moment and shear on the frame beam of other floors and interlayer frame column.

Under the above assumptions, an n-layer frame can be decomposed into n frames, where the ith framework only contains i^{th} beams and columns connected to these beams, and the distal ends of these columns are assumed to be fixed. The bending moment and shear force of the original frame are the superposition of the bending moment and the shear force of these n frames. As shown in Figure 4.5, a four-storey frame is decomposed into layers by the layering method. Obviously, it is easy to compute the Nth frames after decomposition, and can be computed by the moment distribution method.

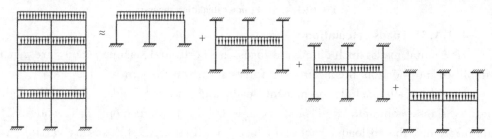

FIGURE 4.5　A schematic diagram for calculating the internal force of a frame structure by a stratified method

In fact, the distal ends of the column are not fixed as shown in Figure 4.5, except for the bottom layer, the constraint in elastic restrained state between the hinge and the fixed. To reflect this situation, the following two assumptions are introduced:

① Except for the bottom layer, the linear stiffness of other columns is multiplied by 0.9;

② The bending moment transfer coefficients of other columns except the bottom column are 1/3.

The moment diagram obtained by the method of delamination shows an imbalance in the bending moment at the joint. In order to improve the accuracy, the unbalanced bending moment can be redistributed once again. The moment distribution is not the distribution of the moment in the sense of stratification, but the other ends of the rod are all elastically restrained except the foundation. Therefore, except for the bottom column, the linear stiffness of the other rods should be multiplied by 0.9.

When calculating hierarchically, as described in the latter 4.3, only the most unfavorable live load arrangements are applied at each level. For the sake of simplicity, live loads can be distributed over all spans at the same time. To reduce the resulting error, multiply the resulting midspan moment by a factor of 1.10 to 1.20.

EXAMPLE 4.1

Condition: The two storey frame is shown in Figure 4.6, and the numbers in brackets are the relative value of the stiffness of members. The load, span and height are also shown in the diagram.

Requirement: using layered method to draw the frame shear diagram.

SOLUTION

(1) Internal force calculation of upper frame

Multiply the relative stiffness of the upper column by 0.9, calculate the distribution coefficient of each node, and write in the brackets of the chart. As shown in Figure 4.6, the value with " * " is the fixed end moment. The transfer coefficient of bending moment is 1/3, and each node is redistributed twice. The process and results are shown in the diagram.

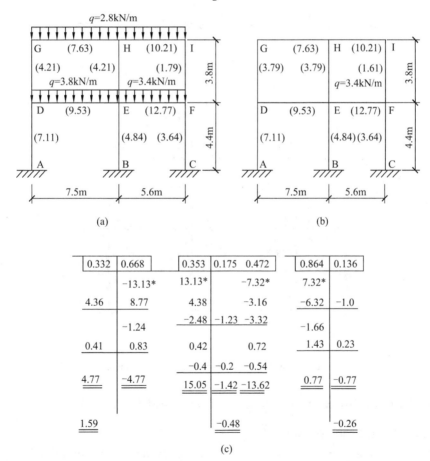

FIGURE 4.6 Calculation of the internal force of the upper frame

(2) Internal force calculation of lower frame

Figure 4.7a shows the internal force calculation process and results of the lower frame. The transfer coefficient of bottom layer is 1/2, and the rest is 1/3.

(3) The last internal force diagram of each frame of the whole frame

The bending moment diagram after superposition of the calculated results is shown in Figure 4.7b. The graph shows that there is an imbalance of forces at the joint. If the difference is large, it can be redistributed.

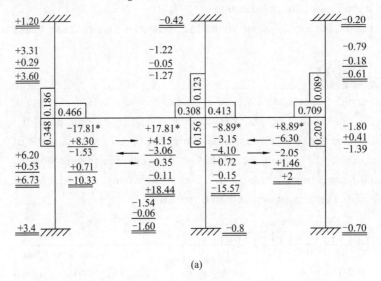

(a)

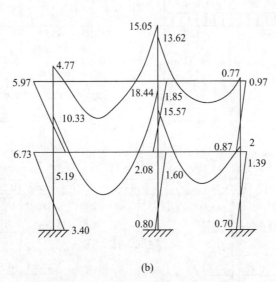

(b)

FIGURE 4.7 Final bending moment diagram

4.2.3 Internal Force Analysis under Horizontal Loads

4.2.3.1 Inflexion point method

Horizontal loads (wind or seismic action) can generally be simplified as horizontal forces acting on the frame joints. The typical moment diagram of the regular frame column under horizontal force of the node is shown in Figure 4.8. The point where the bending moment is zero is called the inflection point. Under the horizontal load of the node, the beam-column node will produce the horizontal displacement and angular displacement. If it is assumed that the line stiffness of the beam is infinite relative to the line stiffness of the column, the node rotation angle is zero when the axial deformation of the column is neglected. Generally when the ratio of the stiffness of the beam to the stiffness of the column is greater than 3, the error caused by the above assumptions will satisfy the accuracy requirements of the engineering design.

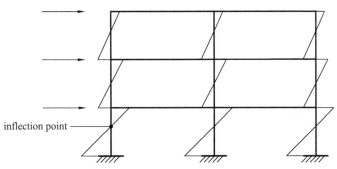

FIGURE 4.8 Bending moment diagram of a frame column under the action of horizontal joint force

Under this assumption, the midpoint bending moment of the column height is zero, that is, the inflection point is at the midpoint of the column height. Considering this feature, the internal force analysis of frame under horizontal load can be simplified, and the analysis method is called the inflection point method.

In engineering design, the inflection point of the bottom column is located at the point 2/3 of the column height from the foundation and the inflexion point of the other column is located at the midpoint of the column height.

It is assumed that the framework has n layers and every layer has n_i columns. The total shear force at i^{th} layer can be calculated according to the equilibrium condition. The shear strength of each column at i^{th} layer are V_{i1}, V_{i2}, $\cdots$, V_{in}, then:

$$V_i = \sum_{k=1}^{n_i} V_{ik} \tag{4-4}$$

Suppose the horizontal displacement of the layer is Δ_i, as there is no horizontal rotation at the two ends of the column, then we have:

$$V_{ik} = \frac{12 i_{ik}}{h_i^2}\Delta_i \tag{4-5}$$

Where, i_{ik} —line stiffness of k column i layer;

h_i —height of i layer column.

Since the stiffness of the beam is infinite and the axial deformation of the beam is zero, so the relative horizontal displacements of the columns at the end of the ith layer are the same (Δu), hence:

$$\Delta u_i = \frac{V_i}{\sum_{k=1}^{n_i} \frac{12 i_{ik}}{h_i^2}} \tag{4-6}$$

Take type (4-5) into type (4-4), the force of column on i layer can be get:

$$V_{ik} = \frac{i_{ik}}{\sum_{k'=1}^{n_i} i_{ik'}} V_i, \; k = 1, \cdots, n_i \tag{4-7}$$

After calculating the shear force of each column, the bending moment of each column can be obtained according to the position of the buckling point of each column.

When the moment of all columns is calculated out, considering the moment equilibrium of each node. For each node, the sum of the moment $\sum M_b$ at the end of the beam can be obtained which equals the sum of the moment at the end of the column. The beam moment at each end of the beam can be calculated by assigning $\sum M_b$ to the line stiffness of the beam connected with the node (i. e., the assigned moment of the beam is proportional to its line stiffness).

EXAMPLE 4.2

Condition: A frame structure and its relative stiffness of members are shown in Figure 4.9.

Requirement: calculate the inner force of the frame by inflection point method and draw the moment diagram.

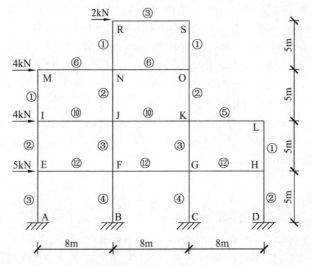

FIGURE 4.9   Layout of frame

SOLUTION

The inflection point of the bottom column is located at $\frac{2}{3}h$ from the bottom of the column, and inflection points of the other columns are at the midpoint of the column height. Cut the column at the inflection point, as shown in Figure 4.10. In order to facilitate the drawing, each layer is cut separately to obtain the shear force and merge into a graph.

The shear force of column is obtained by:

$$V = \frac{d_i}{\sum d_i} \sum P$$

$\sum P$ is the sum of all horizontal forces above the inflection point.

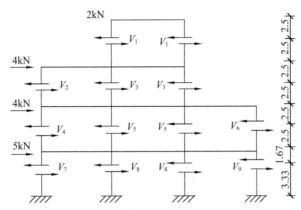

FIGURE 4.10 Shear force of frame

Shear force of top column:
$$V_1 = \frac{1}{1+1} \times 2 = 1\text{kN}$$

Shear force of third layer column:
$$V_2 = \frac{1}{1+2+2} \times (2+4) = 1.2\text{kN}$$
$$V_3 = \frac{2}{1+2+2} \times (2+4) = 2.4\text{kN}$$

Shear force of second layer column:
$$V_4 = \frac{2}{2+3+3+1} \times (2+4+4) = 2.22\text{kN}$$
$$V_5 = \frac{3}{2+3+3+1} \times (2+4+4) = 3.33\text{kN}$$
$$V_6 = \frac{1}{2+3+3+1} \times (2+4+4) = 1.11\text{kN}$$

Shear force of bottom column:
$$V_7 = \frac{3}{3+4+4+2} \times (2+4+4+5) = 3.45\text{kN}$$
$$V_8 = \frac{4}{3+4+4+2} \times (2+4+4+5) = 4.62\text{kN}$$
$$V_9 = \frac{2}{3+4+4+2} \times (2+4+4+5) = 2.31\text{kN}$$

Figure 4.11 is Moment diagram of the frame. Taking node K as an example, the calculation of the bending moment at the column end and the beam end is shown in Figure 4.11.

Column:
$$M_{KD} = V_3 \times 2.4 = 2.4 \times 2.5 = 6\text{kN} \cdot \text{m}$$
$$M_{KG} = V_5 \times 2.5 = 3.3 \times 2.5 = 8.32\text{kN} \cdot \text{m}$$

Unbalanced moment of joints:
$$M_{KD} + M_{KG} = -6 - 8.32 = -14.32\text{kN} \cdot \text{m}$$

Beam:

$$M_{KJ} = \frac{10}{10+5} \times 14.32 = 9.55 \text{kN} \cdot \text{m}, \quad M_{KL} = \frac{5}{10+5} \times 14.32 = 4.77 \text{kN} \cdot \text{m}$$

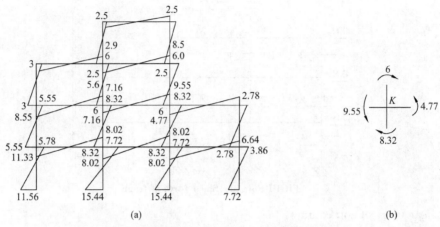

FIGURE 4.11 Moment diagram of a frame

4.2.3.2 D-Value method

The application of the inflection point method is limited by the assumption that stiffness of beam is infinite. In general, the lateral stiffness of the column is also related to the linear stiffness of the beam. The bending point height of the column is also related to the line stiffness ratio of the beam and column, the line stiffness of the upper and lower beams and the floor height of the upper and lower floors. On the basis of the inflexion point method, taking the above factors into account, the D-value method is obtained by modifying the lateral stiffness of the column and the height of the inflexion point. In the D-Value Method, the lateral stiffness of the column is expressed in D, hence the corrected column stiffness D can be expressed as

$$D = \alpha_c \frac{12 i_c}{h_i^2} \tag{4-8}$$

where, i_c, h_i—the stiffness and height of the column;

α_c—the correction coefficient of which that takes into account the elastic constraints of the top and bottom nodes, as shown in Table 4.1.

TABLE 4.1 Formula for calculating lateral stiffness correction coefficient

Position	End column		Middle column		α
	Simple calculation diagram	K	Simple calculation diagram	K	
normal layer	$\begin{array}{c} i_2 \\ i_c \\ i_4 \end{array}$	$K = \dfrac{\sum i_b}{2 i_c}$	$\begin{array}{cc} i_1 & i_2 \\ & i_c \\ i_3 & i_4 \end{array}$	$K = \dfrac{\sum i_b}{2 i_c}$	$\alpha = \dfrac{K}{2+K}$
bottom layer	$\begin{array}{c} i_2 \\ i_c \end{array}$	$K = \dfrac{\sum i_b}{i_c}$	$\begin{array}{cc} i_1 & i_2 \\ & i_c \end{array}$	$K = \dfrac{\sum i_b}{i_c}$	$\alpha = \dfrac{0.5+K}{2+K}$

After obtaining the D value of the lateral stiffness of the column, the kth column, i^{th} layer shear force can be obtained by the method similar to that of the inflexion point method:

$$V_{ik} = \frac{D_{ik}}{\sum_{k=1}^{n_i} D_{ik}} V_i \qquad (4\text{-}9)$$

When the shear of the column is known and the bending moment of the column is required, the bending point of the column should be known. Obviously, the location of the inflection point of the column depends on the ratio of the bending moments at the upper and lower ends of the column. The factors that affect the position of the inflexion point of the column include: the number of layers, the level of the column, the stiffness ratio of the beam and column, and the height variations of the upper and lower layers. The above factors are:

(1) The influence of the beam and column stiffness ratio and the number of layers on the impact of the location of the inflection point

Considering the influence of the beam-column stiffness ratio, the structural layers and gradation on the height of the inflection point, Suppose the line stiffness of the beam, the line stiffness of the column and the height of the storey remain unchanged, the height of the inflection point $y_0 h_i$ of each column of different layer can be calculated according to D. 1.

(2) The influence of the upper and lower cross-beam stiffness ratio impact on the height of the inflection point

If the line stiffness of the upper and the lower cross beam is different, inflection point has a tendency to offset to the transverse beam with smaller rigidity. Therefore, the y_0 must be corrected by an increment y_1 plus one increment. The value of y_1 can be checked in the Appendix D.

(3) The influence of the change of layers on the inflection point. when the height of layer change, the inflection point should go up or down with value of $y_2 h_i$ or $y_3 h_i$, y_2 or y_3 can be found in Appendix D.

For the top column, without considering the correction value y_2, that is, takes $y_2 = 0$. For the bottom column, without considering the correction value y_3, that is, takes $y_3 = 0$.

After the above amendments, the height of the bottom of the column to the inflexion point yh_i is

$$yh_i = (y_0 + y_1 + y_2 + y_3) h_i \qquad (4\text{-}10)$$

After obtaining the lateral stiffness of each column and the height of the inflexion point of each column, the shear force, bending moment and axial force of each column, the bending moment and the shear force of the column beams can be obtained by the inflection point method.

EXAMPLE 4.5

Calculating bending moment diagram of a frame by D-value method (fourth layers).

Condition: The structure of a four-storey frame is shown in Figure 4.12. The beam and column are cast-in-place and the floor is prefabricated. The column cross section size is 400mm × 400mm, and the section size of the top beam is 240mm × 600mm, the section size of floor beam is 240mm × 650mm, the section size of aisle beam is 240mm × 400mm, the concrete grade is C20.

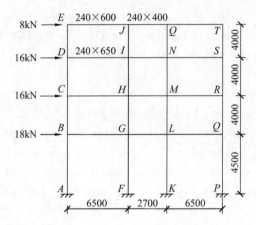

FIGURE 4.12 Layout and load of frame

Requirements: using D value method calculate the internal force of the frame structure under horizontal load.

SOLUTION

(1) Stiffness calculation of beam column line

Component	Moment of inertia of cross section I/mm^4	Linear stiffness $i=\dfrac{EI}{l}/(\text{N}\cdot\text{mm})$	Relative line stiffness
Beam of top layer	$\dfrac{240\times 600^3}{12}=4.32\times 10^9$	$\dfrac{4.32\times 10^9}{6500}E=6.65\times 10^5 E$	0.787
1-3-layer beam	$\dfrac{240\times 650^3}{12}=5.49\times 10^9$	$\dfrac{5.49\times 10^9}{6500}E=8.45\times 10^5 E$	1.000
Bottom beam	$\dfrac{240\times 400^3}{12}=1.28\times 10^9$	$\dfrac{1.28\times 10^9}{2700}E=4.74\times 10^5 E$	0.500
2-4-layer column	$\dfrac{400\times 400^3}{12}=2.13\times 10^9$	$\dfrac{2.13\times 10^9}{4000}E=5.53\times 10^5 E$	0.631
Bottom column		$\dfrac{2.13\times 10^9}{4500}E=4.73\times 10^5 E$	0.500

(2) Calculate the shear value of each column

	Column DE	Column IJ	Column NO	Column ST	$\sum D$
Fourth layers	$\bar{K}=\dfrac{1+0.787}{2\times 0.631}$ $=1.416$ $D=\dfrac{1.416}{2+1.416}\times$ $0.631\times\left(\dfrac{12}{4^2}\right)$ $=0.262\left(\dfrac{12}{4^2}\right)$ $V=\dfrac{0.262}{1.2}$ $=1.75\text{kN}$	$\bar{K}=\dfrac{2\times 0.561+1+0.787}{2\times 0.631}$ $=2.305$ $D=\dfrac{2.305}{2+2.305}\times 0.631\times$ $\left(\dfrac{12}{4^2}\right)=0.338\left(\dfrac{12}{4^2}\right)$ $V=8\times\dfrac{0.338}{1.2}$ $=2.25\text{kN}$	$V=2.25\text{kN}$	$V=1.75\text{kN}$	$\sum D=$ $1.200\left(\dfrac{12}{4^2}\right)$
	Column CD	Column HI	Column MN	Column CD	$\sum D$
Third layer	$\bar{K}=\dfrac{1+1}{2\times 0.631}$ $=1.585$ $D=\dfrac{1.585}{2+1.585}\times$ $0.631\times\left(\dfrac{12}{4^2}\right)$ $=0.279\left(\dfrac{12}{4^2}\right)$ $V=(8+16)\times$ $\dfrac{0.279}{1.256}$ $=5.33\text{kN}$	$\bar{K}=\dfrac{2\times(1+0.561)}{2\times 0.631}$ $=2.474$ $D=\dfrac{2.474}{2+2.474}\times 0.631\times$ $\left(\dfrac{12}{4^2}\right)=0.349\left(\dfrac{12}{4^2}\right)$ $V=(8+16)\times\dfrac{0.349}{1.256}$ $=6.67\text{kN}$	$V=6.67\text{kN}$	$V=5.33\text{kN}$	$\sum D=$ $1.256\left(\dfrac{12}{4^2}\right)$
	Column BC	Column GH	Column LM	Column OR	$\sum D$
Second layer	$D=0.79\left(\dfrac{12}{4^2}\right)$ $V=(8+16+16)\times$ $\dfrac{0.279}{1.256}$ $=8.89\text{kN}$	$D=0.279\left(\dfrac{12}{4^2}\right)$ $V=(8+16+16)\times$ $\dfrac{0.349}{1.256}$ $=11.11\text{kN}$	$V=11.11\text{kN}$	$V=8.89\text{kN}$	$\sum D=$ $1.256\left(\dfrac{12}{4^2}\right)$
	Column AB	Column FG	Column KL	Column PO	$\sum D$
First layer	$\bar{K}=\dfrac{1}{0.561}$ $=1.783$ $D=\dfrac{0.5+1.783}{2+1.783}\times$ $0.560\left(\dfrac{12}{4^2}\right)$ $=0.338\left(\dfrac{12}{4^2}\right)$ $V=(8+16+16+$ $18)\times\dfrac{0.338}{1.446}$ $=13.56\text{kN}$	$\bar{K}=\dfrac{1+0.561}{0.561}$ $=2.873$ $D=\dfrac{0.5+2.873}{2+2.873}\times$ $0.561\left(\dfrac{12}{4^2}\right)$ $=0.385\left(\dfrac{12}{4}\right)$ $V=(8+16+16+18)\times$ $\dfrac{0.385}{1.446}$ $=15.44\text{kN}$	$V=15.44\text{kN}$	$V=13.56\text{kN}$	$\sum D=$ $1.446\left(\dfrac{12}{4^2}\right)$

(3) Calculate the height of inflection point of each column

According to the total number of layers M, column layer n, stiffness ratio of beam to column, and the standard inflection point height, the coefficient y_0 is obtained by referring to the table D.1. According to the transverse beam rigidity ratio, refer to table to obtain the correction value y_1. According to the height change of the upper and lower layers, refer to table to obtain correction value y_2, y_3. The height of the inflection point is given by $y_h = (y_0 + y_1 + y_2 + y_3)h_0$.

	Column *DE*	Column *U*	Column *NO*	Column *ST*
Fourth layers	$\bar{K}=1.416$ $\alpha_1=\dfrac{0.787}{1}=0.787$ $\alpha_3=1$ $y_0=0.37$ $y_1=0$ $y_3=0$ $y=0.37+0+0=0.37$	$\bar{K}=2.305$ $\alpha_1=\dfrac{0.787+0.561}{1+0.561}=0.846$ $\alpha_3=1$ $y_0=0.42$ $y_1=0$ $y_3=0$ $y=0.42+0+0=0.37$	$y=0.42$	$y=0.37$
	Column *CD*	Column *HI*	Column *MN*	Column *CD*
Third layer	$\bar{K}=1.85$ $\alpha_1=1$ $\alpha_2=1$ $\alpha_3=1$ $y_0=0.45$ $y_1=0$ $y_2=0$ $y_3=0$ $y=0.45+0+0=0.45$	$\bar{K}=1.85$ $\alpha_1=1$ $\alpha_2=1$ $\alpha_3=1$ $y_0=0.47$ $y_1=0$ $y_2=0$ $y_3=0$ $y=0.47+0+0=0.47$	$y=0.47$	$y=0.45$
	Column *BC*	Column *GH*	Column *LM*	Column *OR*
Second layer	$\bar{K}=1.85$ $\alpha_1=1$ $\alpha_2=1$ $\alpha_3=1$ $y_0=0.45$ $y_1=0$ $y_2=0$ $y_3=0$ $y=0.45+0+0=0.45$	$\bar{K}=1.85$ $\alpha_1=1$ $\alpha_2=1$ $\alpha_3=1$ $y_0=0.47$ $y_1=0$ $y_2=0$ $y_3=0$ $y=0.47+0+0=0.47$	$y=0.47$	$y=0.45$
	Column *AB*	Column *FG*	Column *KL*	Column *PO*
First layer	$\bar{K}=1.783$ $\alpha_2=0.889$ $y_0=0.5$ $y_2=0$ $y=0.55$	$\bar{K}=1.783$ $\alpha_2=0.889$ $y_0=0.5$ $y_2=0$ $y=0.55$	$y=0.55$	$y=0.55$

(4) The Bending moment of the upper and lower ends of the column

It is calculated by the following formula: upper column $M_{上} = V(1-y)h$, lower column $M_{下} = Vyh$. The bending moment of each beam is calculated by the equilibrium condition of the node and the line stiffness ratio of the beam.

4.2.4 Calculation of Displacement and the Limit Value of Frames

The lateral displacement of the frame structure is mainly caused by the horizontal load. Therefore, the lateral displacement is usually calculated only under horizontal load. The deformation of the frame structure under horizontal load consists of two parts: the total shear deformation and the overall bending deformation of the beam and column. The overall shear deformation is caused by the bending deformation of the beam and column, and its lateral displacement curve is consistent with the shear deformation of the cantilever as shown in Figure 4.13. The overall bending deformation is caused by the axial deformation of the frame column, and its lateral displacement curve is consistent with the bending deformation of the cantilever beam, as shown in Figure 4.14.

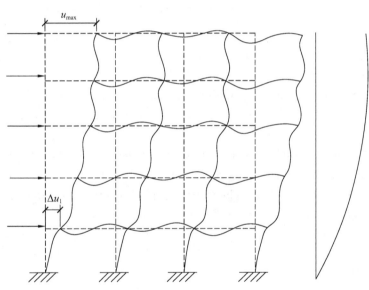

FIGURE 4.13 The overall shear deformation of the frame (caused by the bending of beam and column)

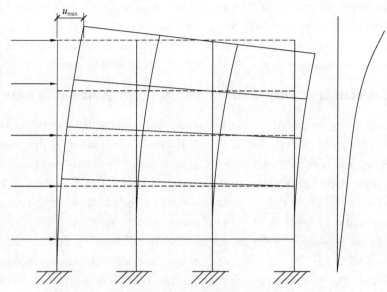

FIGURE 4.14 The overall bending deformation of the frame
(caused by the axial deformation of the beam and column)

For the conventional frame structure, the lateral displacement curve is dominated by the overall shear deformation, so only the lateral displacement due to the bending of the beam column is considered. When the overall aspect ratio of the structure increases, the ratio of the overall bending deformation increases. When the total height $H>50$m or aspect ratio $H/B>4$, it is necessary to consider the lateral displacement caused by the axial deformation of the column.

4.2.4.1 Lateral displacement caused by bending beams and columns

The lateral displacement caused by the bending deformation of the frame beam column is shown in Figure 4.7. The lateral displacement of the vertex μ_M is the sum of the lateral displacement between the layers,

$$\mu_M = \sum_{j=1}^{n} \Delta\mu_j \tag{4-11}$$

and

$$\Delta u_j = \frac{V_{jk}}{D_{jk}} = \frac{V_{Fj}}{\sum_{k=1}^{m} D_{jk}} \tag{4-12}$$

4.2.4.2 Lateral displacement caused by axial deformation of the column

The lateral displacement caused by axial deformation of the frame column is shown in Figure 4.8. When the frame is under horizontal load, the column produces axial tension on one side and axial compression on the other side. The axial force of the outer column is large and the axial force of the inner column is small.

For simplicity, the axial force of the inner column is ignored and the axial force of the outer column is approximated as:

$$N = \pm \frac{M}{B} \tag{4-13}$$

Where, M—the bending moment at the point of caused by horizontal force;

B—the distance between the outer columns.

The horizontal displacement of the vertex of the frame can be obtained by the unit load method

$$u_N = \sum \int_0^H \frac{N_1 N}{E_c A} dZ \qquad (4\text{-}14)$$

where, N_1—the axial force of the side column caused by a unit horizontal force acting on top of the frame;

N—column axial force under horizontal external load;

A—sectional area of column;

E_c—elastic modulus of a column.

4.2.4.3 The limit of lateral displacement

In addition to ensuring that the deflection of the beam does not exceed the specified value, the lateral displacement of the structure should also be checked. For checking the lateral displacement of the structure, the inter story displacements should meet the following requirements:

$$\Delta u = /h \leqslant [\Delta u/h] \qquad (4\text{-}15)$$

where, h—storey height of structure;

Δu—inter-storey displacement calculated by elastic calculation method;

$[\Delta u/h]$—the interlayer displacement limit of the frame structure, it takes the value of Table 4.2.

TABLE 4.2   The interlayer displacement angle limit of frame structure

Height/m	$[\Delta u/h]$
$H \leqslant 150$	1/550
$150 < H < 250$	linear interpolation
$H \geqslant 250$	1/500

Note: the maximum displacement between floors is calculated by the maximum horizontal displacement difference method without deducting the overall bending deformation.

EAMPLE 4.4

Calculation of the frame lateral displacement.

Requirements: calculate the frame lateral displacement of the frame in Figure 4.15.

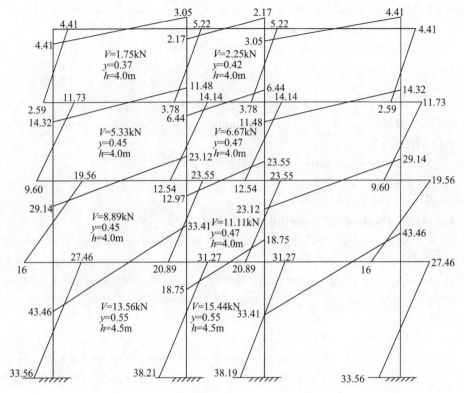

FIGURE 4.15 Frame bending moment diagram

SOLUTION

The concrete strength grade is C20, elastic modulus $E = 25.5 \times 10^3 \text{N/mm}^2$
Total frame height $H = 16.5\text{m} < 50\text{m}$, width $B = 15.7\text{m}$
$\dfrac{H}{B} = \dfrac{16.5}{15.7} = 1.051 < 4$, the lateral displacement caused by axial deformation of the column can be neglected. Only the lateral displacement caused by the bending deformation of the beam is calculated.

$$\Delta i = \frac{V_i}{\sum D_i}$$

(1) Calculating the lateral stiffness of each storey $\sum D_i$, this should be the absolute value of the lateral stiffness
$i = 8.45 \times 10^5 \times 25.5 \times 10^3 = 215.48 \times 10^5 \text{N} \cdot \text{mm}$
The absolute value of the lateral stiffness of each layer are:

top layer: $D_4 = 1.2 i \left(\dfrac{12}{H_4^2}\right) = 1.2 \times 215.48 \times 10^5 \times \left(\dfrac{12}{4000^2}\right) = 1.94 \times 10^4 \text{N/mm}$

third layer: $D_3 = 1.256 i \left(\dfrac{12}{H_3^2}\right) = 1.256 \times 215.48 \times 10^5 \times \left(\dfrac{12}{4000^2}\right) = 2.03 \times 10^4 \text{N/mm}$

second floor: $D_2 = 1.256 i \left(\dfrac{12}{H_2^2}\right) = 1.256 \times 215.48 \times 10^5 \times \left(\dfrac{12}{4000^2}\right) = 2.03 \times 10^4 \text{N/mm}$

first floor: $D_1 = 1.446 i \left(\dfrac{12}{H_1^2}\right) = 1.446 \times 215.48 \times 10^5 \times \left(\dfrac{12}{4000^2}\right) = 1.85 \times 10^4 \text{N/mm}$

(2) Calculate the inter-storey displacement

Shear force of each layer $V_4 = 8\text{kN}$; inter-storey lateral displacement $\Delta_4 = \dfrac{8 \times 10^3}{1.94 \times 10^4} = 0.412\text{mm}$

$V_3 = 8 + 16 = 24\text{kN}$; $\Delta_2 = \dfrac{24 \times 10^3}{2.03 \times 10^4} = 1.182\text{mm}$

$V_2 = 8 + 16 + 16 = 40\text{kN}$; $\Delta_2 = \dfrac{40 \times 10^3}{2.03 \times 10^4} = 1.97\text{mm}$

$V_1 = 8 + 16 + 16 + 18 = 58\text{kN}$; $\Delta_1 = \dfrac{58 \times 10^3}{1.85 \times 10^4} = 3.14\text{mm}$

(3) Calculate top lateral displacement

$\Delta_V = \sum\limits_{i=1}^{4} \Delta_i = 0.412 + 1.182 + 1.97 + 3.14 = 6.7\text{mm}$

4.3 The Most Unfavourable Position of Load Combination

4.3.1 Control Sections

The bending moment of the frame column varies linearly along the column height, and the maximum bending moment is at both ends of the column. Therefore, the upper and lower ends of each column can be taken as the control section. For the frame beam, under the combined action of horizontal force and vertical load, the shear force changes linearly along the beam axis, while the bending moment changes parabolically (referring to the vertical load distributed). The maximum positive bending moment and the maximum shear is often at both ends of the cross section. Therefore, besides the beam ends, the cross-section of the maximum positive bending moment should also be taken as the control section. For simplicity, the maximum positive moment control section is not determined by the method of extreme value, the mid-section of the beam is taken as the control section directly.

The beam bending moment and shear force obtained from internal force analysis are all internal force values at the axis, but the internal force at the end of the member should be used instead of the internal force at the axis during the design of the member or the section reinforcement. The internal forces at the end of the column cross-section of the beam can be calculated. The internal forces at the end cross sections of the column can be calculated similarly (in order to ensure the safety of the column, the internal force at the intersection of the beam column axis is usually used).

4.3.2 Load Effect Combination

In structural design, the most unfavorable condition of the simultaneous action of different load must be taken into account. The variable loads on a multi-storey frame structure consist of two items, namely, wind load and floor variable load. There are three combinations:

① 1.2 internal force generated by the permanent load standard value +1.4 internal force generated by floor variable load standard value +1.4×0.6 internal force generated by wind load standard value.

② 1.2 internal force generated by the permanent load standard value +1.4 internal force generated by wind load standard value +1.4×0.7 internal force generated by floor variable load standard value.

③ 1.35 internal force generated by the permanent load standard value +1.4×0.6 wind load standard value +1.4×0.7 internal force generated by floor variable load standard value.

For seismic fortification zone, horizontal seismic action combination should be taken into account: 1.2×internal force generated by the standard value of gravity load +1.3× internal force generated by the standard value of the horizontal seismic action.

When designing the component of the frame structure, the most unfavorable internal force of each component must be calculated out. For a control section of a component, there may be several combinations of the most unfavorable internal forces. For example, for the beam end section of concrete frame, in order to calculate the reinforcement at the top of the beam, the maximum negative moment of the section must be calculated; in order to determine the reinforcement at the bottom of the beam, the maximum positive moment of the section must be calculated; The maximum shear of the section is used to calculate the shear capacity of the beam ends. In general, not all loads acting at the same time which results in the maximum bending moment of the section, some loads produces the maximum positive moment of a control section, but some other loads induce the maximum negative moment of the cross section.

There are many load combinations of the most unfavorable internal forces. For the frame beam section, the most unfavorable internal forces are the maximum negative bending moment and the maximum shear force, but also the possible positive bending moments. For the mid-section of the frame beam, the most unfavorable internal force is the maximum positive bending moment or negative bending moment.

For the upper and lower sections of the frame columns, under large eccentric compression situation, it is more unfavorable when the moment M becomes more large. Under small eccentric compression situation, it is more unfavorable when the axial force N becomes more large. Therefore, the most unfavorable internal force of the general frame structure is summarized as:

(1) beam end section: $+M_{max}$, $-M_{max}$, V_{max}

(2) middle section of beam: $+M_{max}$, $-M_{max}$

(3) column end section: ① $|M_{max}|$ and N, V

② N_{max} and M, V

③ M_{min} and M, V

In some cases, though the value of internal forces is not the maximum or minimum it may also be the most unfavorable. For example, for a small eccentric cross-section of a concrete frame column, when N is not the maximum, but the corresponding M is relatively large. In this case, it needs more reinforcement which has the most unfavorable internal force, therefore attention should be paid when designing.

4.3.3 Most Unfavorable Position under Vertical Variable Loads

In order to obtain the most unfavorable internal force of the control section, the most unfavorable arrangement of variable loads should be considered. The most unfavorable position of a variable load can be determined by the influence line.

For the multi-storey frame shown in Figure 4.9, to get the influence line of the most unfavorable effect of the maximum positive moment load of the cross-section at midspan C of the span AB beam, the influence line M_C should be made first. Remove the corresponding constraint (i.e. change the point C to a hinge) to make the structure produce a unit virtual displacement in the positive direction of the binding force, thus the virtual displacement diagram of the entire structure can be obtained, as shown in Figure 4.16a. In order to obtain the maximum positive bending moment of beam span AB, it is necessary to arrange a variable load between every span that produces a positive virtual displacement, which forms a checkerboard spacing arrangement as shown in Figure 4.16b. It can be seen when internal moment of the span AB reaches the maximum value under the most unfavorable arrangement of various load, the internal moments of other span with variable load also get maximum. As a result, the maximum positive internal moment of all the frames beams can be obtained.

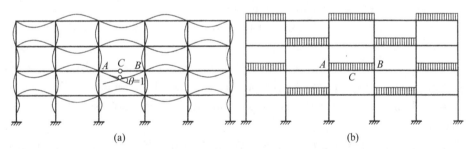

FIGURE 4.16 The most unfavorable variable load arrangement of the maximum bending moment in a multi-layer frame beam

The most unfavorable arrangement of variable load for the maximum negative bending moment at the beam end or the maximum moment at the column end can also be obtained by the above method. However, during the solution, it is sometimes difficult to determine when the direction of the virtual displacement for frame nodes that are distant from the calculated section. In this case, the influence of the load from the calculated section on the internal force of the calculated section is negligible.

Obviously, the most unfavorable situation of the maximum axial force of the column is the variable load are applied on the beams that are connected to the column in the above layers.

When the internal force under vertical load is calculated by the stratification method, only the unfavorable arrangement of variable load in the layer needs to be considered. The arrangement method is the same as the most unfavorable arrangement method of variable load of continuous beam. For the moment of column ends, it is only necessary to consider the unfavorable arrangement of variable loads on the upper and lower adjacent columns. If the span is not very large, the variable load can be applied to the open frame one by one, and the internal forces of the structure can be calculated individually, then the most unfavorable internal force can be calculated out according to the unfavorable combination of the section.

When the internal force generated by the vertical variable load is less than the internal force generated by the permanent load and the horizontal load, the vertical variable load can be simultaneously applied to all the frame beams without consideration of the most unfavorable arrangement of the variable load. The internal force obtained at the support by this method is very similar to the internal forces according to the method of most unfavorable load position. However, the calculated midspan moment of the beam is smaller than that of the most unfavorable load position method, so the moment in the beam midspan should be multiplied by the coefficient of 1.1~1.2.

4.3.4 Moment Amplitude Modulation of Beam End

In accordance with the reasonable failure of the frame structure, plastic hinges are allowed to occur at the beam ends; for assembled or assembled monolithic frames, the nodes are not absolutely rigid; the internal force diagram of the frame is calculated according to the elastic theory. The beam end bending moment will be less than the actual value of the elastic method calculation.

For reinforced concrete frames, the redistribution of the plastic deformation at the end of the frame beam can be considered under the action of vertical loads. During the design of the frame structure, the beam bending moment is generally modulated, i.e., reduce the negative bending moment of the beam end, thus the amount of reinforcement at the top of the beam near the node is reduced, so as to save steel and facilitate the construction.

When a frame beam AB is under vertical load, the maximum negative bending moment at the end of beam is M_{A0} and M_{B0} respectively and the maximum positive bending moment of beam is M_{C0}, then the moment at the end of the beam after amplitude modulation is

$$M_A = \beta M_{A0}$$
$$M_B = \beta M_{B0} \tag{4-16}$$

where, β—beam moment modulation coefficient (for the cast-in-place frame, $\beta = 0.8 \sim 0.9$; for the assembled integral type frame, the welding joint is not strong or the node area of concrete perfusion is less dense, nodes are prone to deform and do not reach the absolute rigidity, therefore, the moment modulation coefficient are allowed to take lower value, usually $\beta = 0.7 \sim 0.8$).

After moment adjusting of beam ends, the midspan bending moment will be increased under the corresponding load, and the moment in the span of beam should be adjusted according to the balance condition

$$M_C = M_0 + \frac{M_A + M_B}{2} \quad (4\text{-}17)$$

where, M_0—midspan bending moment calculated by simply supported beam.

Because of the vertical load, the beam end supports is generally negative bending moment, so the above equation can be written as

$$M_C = M_0 - \left| \frac{M_A + M_B}{2} \right| \quad (4\text{-}18)$$

Moreover, in the design of section, in order to ensure that the reinforcement at the bottom of the midspan frame beam is not relatively small, the design value of the positive moment of the midspan section should not be less than 50% of the design value of the midspan bending moment calculated by the simple beam under the vertical load, that is:

$$M_C \geqslant \frac{M_0}{2} \quad (4\text{-}19)$$

Moment amplitude modulation of internal force only considers vertical load, the moment generated by horizontal load does not participate in the amplitude modulation. Therefore, the moment amplitude modulation should be carried out before the internal force combination.

4.4 Design of Frame Structural Components

4.4.1 Design of Frame Beam and Column

The calculation of frame beam cross-section includes: the calculation of bearing capacity of the cross-section under ultimate limit state; calculation of crack width and deflection under service ability limit state. For non-seismic conditions, the cross-section calculation method and construction requirements of reinforced concrete frame beam are the same as those in Chapter 2.

The calculation of the section of frame column includes: the bearing capacity of the normal section and the inclined section under ultimate limit state; the crack width under the service ability limit state (the width of the crack donit need be calculated for the eccentric bearing column when $e_0 h_0 \leqslant 0.55$).

The diagonal section bearing capacity of the frame column is calculated as follows

$$V \leqslant \frac{1.75}{\lambda + 1} f_t b h_0 + f_{yv} \frac{A_{sv}}{s} h_0 + 0.07N \quad (4\text{-}20)$$

where, λ—the shear span ratio of the calculated section. When the inflection point is in the high storey range, the shear span ratio can take $\lambda = H_n/(2h_0)$, and H_n is the column clear height, $1 \leqslant \lambda \leqslant 3$;

N—design value of axial force corresponding to design value of shear force, when $N \geqslant 0.3 f_c A$, take $N = 0.3 f_c A$.

4.4.2 Design of Frame Joints

Joints design is a very important part of the frame structure design. Under the condition of no anti-seismic fortification requirement, the current node design theory is very difficult to guarantee the bearing capacity of the node from the calculation, so it is necessary to take the appropriate structural measures to ensure the bearing capacity of the joint.

Failure caused by component failure may be localized, and failure caused by node failure is often large-scale, hence the consequences are serious. The importance of nodes is obvious.

The joint design should ensure the safety, reliability, economy and easy construction. For the precast-monolithic frame nodes, to ensure the integrity of the joints, node force, simple structure, convenient installation and easy adjustment, the component connection can bear some or all of the design load, hence the superstructure can be installed in time.

(1) General requirements

The concrete strength grade of the frame joint area is the same as the concrete strength grade of the cast-in-place frame columns. The concrete strength grade requires higher than that of prefabricated structures.

The joint's cross-section size is generally as the same as the column. For the top edge joint, the beam cross-section size should meet the following requirements

$$0.35\beta_c f_c b_b h_0 \geqslant A_s f_y \tag{4-21}$$

where, A_s—the longitudinal reinforcement of the beam at the top end;

b_b, h_0—the width of the web and the effective height of the section;

β_c—the influence coefficient of concrete strength; when the concrete strength grade is equal or less than C50, $\beta_c = 1.0$; when the concrete strength grade is C80, $\beta_c = 0.8$; If the concrete strength grade is between the vales, the linear interpolation method can be used to determine the coefficient value.

The horizontal stirrup should be installed in the joint, and its requirements should be the same as that of the column, but the spacing should not be greater than 250mm. When the top end joint has a lap joint of longitudinal reinforcements of beam up side and column outside, the horizontal stirrups in the node should meet the requirements for setting the stirrups in the overlapping region of the longitudinal bars.

(2) Connection of longitudinal bars of column and anchorage in the joint area

In order to facilitate the construction, the column longitudinal reinforcement connection joints are generally located at the floor. The connecting joints shall be staggered with each other, in the same connecting section the joint area percentage of longitudinal tensile steel bars should not exceed 50%.

The length of the connecting section is determined according to the following rules: The lap length is 1.3 times banding lap, and the lap length l_l is related to the joint area. When the joint area percentage is less than 25%, the length is 1.2 l_a (l_a is the anchorage length of longitudinal tensile reinforcement), when the joint area is 50%, it takes 1.4 l_a which is equivalent to 35d (d is the maximum diameter of the longitudinal reinforcement) for the mechanical and welded joints. The stirrup spacing within the range of the lap length of the longitudinal bearing force reinforcement: when the steel bar is in tension, the stirrup spacing should be no more than 5 times the smaller diameter of longitudinal reinforcement, and not greater than 100mm; when the steel bar is compressed, the stirrup spacing is not more than 10 times the smaller diameter of longitudinal reinforcement, and no more than 200mm, as shown in Figure 4.17.

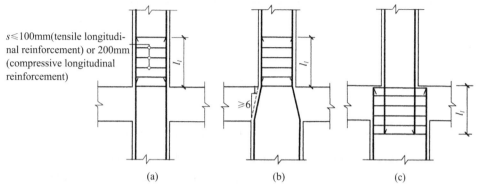

FIGURE 4.17 Lap joint of column longitudinal reinforcement

The column longitudinal reinforcement should be anchored into the top node and its length from the bottom of the beam to the top of the column should not be less than the anchorage length, as shown in Figure 4.18a; when the top beam height is less than l_a, it can be extended to the top of the column and then bent inward horizontally, the length of the vertical end before bending should not be less than 0.5 l_{ab} (which is the basic anchorage length of the tensioned steel), the length of the horizontal section after bending should not be less than $12d$, as shown in Figure 4.18b. When the top layer is a cast-in-place concrete slab and the thickness is not less than 80mm, the column longitudinal reinforcement can also be bent outward horizontally as shown in Figure 4.18c.

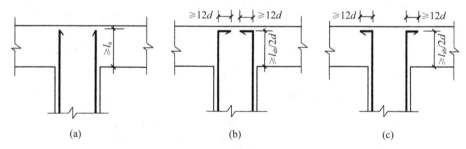

FIGURE 4.18 Anchorage of column longitudinal bar in top floor

(3) Anchorage of beam longitudinal reinforcement in the joint area

The longitudinal bars at the upper part of the frame beam are generally continuous at the middle nodes. When anchoring is required at the node, the length of the extending node should not be less than l_a, and should be extended over the middle center line $5d$ as shown in Figure 4.19a. When the column size is not sufficient, the upper longitudinal reinforcement can be extend to the opposite side of the node and bent downwards. The length of the horizontal section before bending is not less than 0.4 l_{ab} and the vertical section of bending is not less than $15d$, as shown in Figure 4.19b.

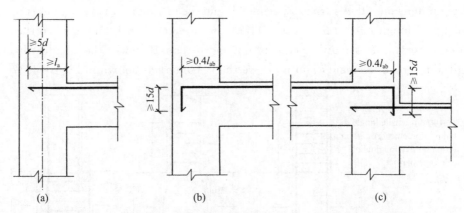

FIGURE 4.19 Anchorage of longitudinal steel bar on the upper part of beam

The lower longitudinal reinforcement of frame beam: When the reinforced strength is not employed, or its compressive strength is used, anchorage is handled by the main beam of slab structure; When the tensile strength of the reinforcement is used (i. e. the lower part of the pedestal is pulled), the length of the extension joint is not less than l_a, as shown in Figure 4.20a; When the pillar section size is insufficient, the lower longitudinal reinforcement can be bent upwards and the horizontal section length is not less than 0.4 l_{ab}, the bending of the vertical section is not less than $15d$, as shown in Figure 4.20b,c.

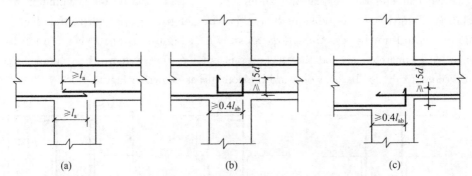

FIGURE 4.20 Anchorage of longitudinal steel bar in the lower part of beam

(4) The overlap of longitudinal reinforcement of up beam and top edge column

There are two schemes of the overlap: One is that the longitudinal reinforcement of column extends into lateral beam, and the lap length is not less than 1.5l_{ab}; for the column angle bar which cannot extend outside the scope of the beam width, reinforcement bar can be inserted into the column iuner side and bend down ward the downward bend, and the bending length can't be less than $8d$, but the number should not exceed 35% of the total lateral longitudinal reinforcement, as shown in Figure 4.21a. Another scheme is that the longitudinal reinforcement of lateral column extends to the top of column, and the upper longitudinal reinforcement of beam extends into the column. The lap length of the vertical section can't be less than 1.7l_{ab}, as shown in Figure 4.21b.

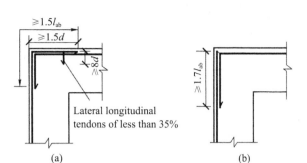

FIGURE 4.21 Lapped bars of top layer joint

4.4.3 Frame Construction Requirements

(1) The construction requirements of cast-in-place frame

The construction requirements of cast-in-place frame structure should not only make the various sections of the components resist the design internal forces effectively, but also have reasonable strength reservation. Thus requires the steel bars are lapped and anchored reasonably to meet the requirements of the minimum reinforcement ratio. Some details have been described before. In addition, the following requirements should also be met.

① At the midspan of the beam, at least $2\phi 12$ steel bars shall be required to lap with the negative bending reinforcement of the beam support, and the lap length shall be no less than $1.2l_a$ of the minimum anchorage length, and also the following shall be included.

② At the lower part of the beam support, at least two longitudinal rebars shall be extended into the column.

③ The spacing of longitudinal steel bars in columns shall not be greater than 300mm, and the net spacing shall not be less than 50mm.

④ The connection of the reinforcing steel bars should be provided in the less stressed area, less joints should be installed on the same bearing-force steel bar. For the important components and key transmission parts of the structure, the longitudinal steel bars should not set with connecting joints.

⑤ When the diameter of the tensile steel bar is greater than 25mm and the diameter of the compression bar is greater than 28mm, the welded joint or mechanical connection shall be adopted.

⑥ The lap length of the longitudinal reinforcing bar in the column shall not be less than $1.2l_a$; the distance between adjacent joints shall not be less than 500mm for welding and not less than 600mm for lap length; the minimum distance from the column to the lower end of the joint shall not be less than the length of the long side of column section, and must be set above the floor level.

⑦ When the longitudinal reinforcement of a column is a lap joint, the percentage of the loaded reinforcement with joints of the total cross-sectional area of the loaded steel reinforcement should not be more than 25% in the tension zone and not more than 50%

in the compression zone. For Column longitudinal welded steel joints in columns, it should not be more than 50% in the tension zone, but no limited in compressive zone.

⑧ The reinforcement ratio of all longitudinal steel bars shall not be greater than 3%, must be less than 5%, and not less than 0.5%.

In addition, the beam column section size and stirrup configuration also have corresponding requirements, readers can refer to the relevant specifications.

(2) The arrangement of beam column joint of prefabricated and prefabricated-monolithic frame

The determination of prefabricated units of fabricated and assembled monolithic frames should make the structure have reasonable loading, simple structure and convenient construction, and consider the feasibility of production, transportation and lifting of prefabricated units. Prefabricated units are usually divided into the following forms:

① Single beam and short column type. The beam and column can be divided into single members by span and height individually, as shown in Figure 4.22a. This model has the advantages such as small components, convenient manufacture, stacking, transportation and lifting. The disadvantage is that the joint is large in number located at the frame. Because the joint is the largest part of the internal force, the scheme is not conducive to bear load of structure.

② Single beam and long column type. The beam is divided by span, and each column is made of two or several layers as a prefabricated component, as shown in Figure 4.22b. In this way, the number of joints and hoisting can be reduced, and the integrity of the building can be improved. Its disadvantage is that it is difficult to lift and transport columns, and the reinforcing bars in the column are often increased due to the lifting and transportation requirements.

③ Frame type. The entire framework is divided into a number of small frames, as shown in Figures 4.22c and d. A small frame may take the form of an H shape or cross shape, etc. The joint position can be at the frame node, or at the middle point of beam with less span and midpoint of the column. This scheme can reduce the number of prefabricated components and hoisting, improve the integrity of the building. The disadvantage is that the prefabricated parts are large and complex, so it is difficult to be produced, transported and hoisted.

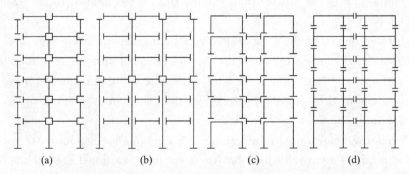

FIGURE 4.22 Construction and division of prefabricated and prefabricated-monolithic

(3) The connection construction of frame beam and precast slab

Prefabricated floors are often groove plate and cored slab. In order to make the floor structure have good integrity, building reinforcement are set up necessarily between the gap, and pouring with fine stone concrete, or casting concrete on the prefabricated plate with grade no less than C20 to form a laminated floor. Its thickness cannot be less than 40mm and two-way reinforced with $\phi 4@150$mm or $\phi 6@250$mm mesh should be used. When it is fabricated monolithic floor, the length of the anchor bar extending to the end floor cannot be less than 100mm, and the minimum length of precast slab supported by the wall or beam is no less than 30mm as shown in Figure 4.23. In order to ensure the reliability during the installation phase, the support length on the beam should be larger than 55mm.

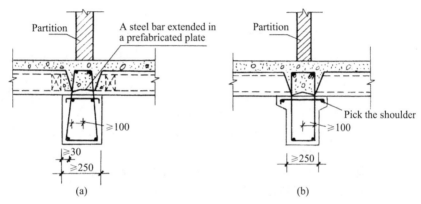

FIGURE 4.23 Connection of prefabricated plate and laminated beam

(4) The construction requirements of infilled walls

Masonry walls are often used in buildings when the position of infilled walls are relatively fixed. At this time, the gap between the upper part of the masonry-infilled wall and the bottom of the frame must be plugged with block materials. There are two ways to connect the wall with the frame column: One is to leave a gap between the column and the wall, and connect flexibly by reinforcement. The calculation does not take into account the influence of the infilled wall on the lateral resistance capacity of the frame. The other one is the rigid connection that the column and the wall are connected tightly. The infill walls, which are rigidly connected to the column, act as diagonal bar (Figure 4.24) when the frame is laterally deformed, thus increase the frame's lateral resistance capacity.

The best material of frame infilled or partition walls is lightweight wall panels and must be securely attached to the frame. When using masonry infilled wall, it should be tied with column by $2\phi 6$ rebars along the height every few layer blocks at the connection of column and wall. The tie rebar should be rigidly anchored into the column, and extend appropriate length into the infill wall as recommended in the specifications.

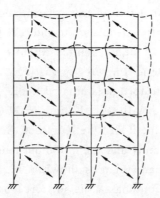

FIGURE 4.24 The effect of the infilled wall as diagonal struts

References

4.1 Zhu Bolong. The principle of concrete structure design [M]. Shanghai: Tongji University Press, 1999.

4.2 Tianjin University, Tongji University, Southeast University. The concrete structure[M]. Beijing: China Construction Industry Press, 2008.

4.3 Zhou Wanghua. Superimposed structure of modern concrete[M]. Beijing: China Construction Industry Press, 1998.

4.4 Zhou Qijin. Handbook for construction of concrete structure[M]. Beijing: China Construction Industry Press, 1994.

Questions

4.1 What kinds of load should be considered in the frame design and describe how the wind load is calculated?

4.2 Describe how the calculation diagram of the frame structure is determined?

4.3 A 10 storey cast-in-place concrete frame structure is shown in Figure 4.25, the total structure height is 32 m. It is located at the 8-degree seismic area, the design seismic group is the first-class site II, the frame's seismic grade is two and the box type basement is the first class. It can be used as the fixing end of the superstructure. The first floor of a side frame column CA, symmetrical reinforcement, height 4.4m. In a load utility combination, the standard values of internal forces produced by the load and the earthquake action at the bottom of the column are as follows:

Permanent load: $M_{GK}=-20\text{kN}\cdot\text{m}$; $N_{GK}=3000\text{kN}$

Floor live load: $M_{QK}=-10\text{kN}\cdot\text{m}$; $N_{QK}=500\text{kN}$

Earthquake action: $M_{Ehk}=\pm 266\text{kN}\cdot\text{m}$; $N_{Ehk}=\pm 910\text{kN}$

Calculate:

(1) The design value of the maximum combined axial force of the bottom section of the column (kN);

(2) The column bottom section of the maximum combined moment design value (kN·m).

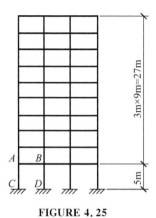

FIGURE 4.25

4.4 A 12 storey residential building, built at the site of fortification intensity of 7 degrees and class II, has a height of 2.9m and a height difference of 900mm. Using cast-in-place reinforced concrete frame structure system, shear ratio of two storey frame columns $\lambda=1.8$. Calculate the limit value of axial compression ratio of the column under the combination of vertical load and seismic load u_N.

4.5 The dimensions of a frame structure is shown in Figure 4.26, $H=17$m, $H/B=1.0$, $\omega_0=0.6\text{kN/m}^2$, B class surface roughness, the windward surface $\mu_s=0.8$ and the leeward surface $\mu_s=-0.5$, the calculation unit of the frame is 6m. Calculate the wind force on the windward and leeward surfaces of the frame structure.

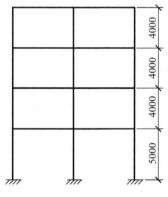

FIGURE 4.26

Chapter 5

Application of Software

5.1 Introduction of Softwares

Structural design software is an important tool for structural engineers when they make complex structural design. It is also a necessary tool to ensure the quality and efficiency of structural design. Hence, we will give a comprehensive introduction to structural design software, especially ETABS.

5.1.1 Introduction of Chinese Softwares

In China, the commonly used structural design softwares are PKPM, YJK and GS.

PKPM is an engineering management software developed by the Architectural Engineering Software Institute of the Chinese Academy of Architectural Sciences. As the name shows, PKPM contains two parts of PK and PM. PM is used to set up models and PK is used to perform analysis. Before 1988, the calculation was based on the input of the original data, by filling in a text file and calculating it by a small program. Today, PMCAD module can be used to input information of storeys and set up 3D models by combination. TAT is used for multi-storey building structures and high-rise building structures, Dynamic time history analysis module, TAT-D is used for high-rise building structures. The Finite element module FWQ help to calculate the frame and shear wall structure and calculate the reinforcement. The Spatial finite element analysis and design module STAWE is also used to design multi-storey and high-rise building structures (Ref. 1.1). The software interface of PKPM is shown in Figure 5.1.

FIGURE 5.1 PKPM software（Ref. 1. 2）

YJK structural calculation software is developed by YJK Building Software Company. YJK structural calculation software was released in 2011 in Beijing. YJK structural calculation software is mainly for space finite element analysis and design of multi-storey high building structures. It is suitable for frame, frame-shear, shear-wall, tube, mixed or steel structures. It uses spatial bar element to simulate the beam, column and support members. The wall element includes shell and rigid-plate element. There are various kinds of elastic plate models for simulation of the slab. One of the good features of YJK is that it provides an friendly and convenient (Ref. 1. 3). The software interface of YJK is shown in Figure 5.2.

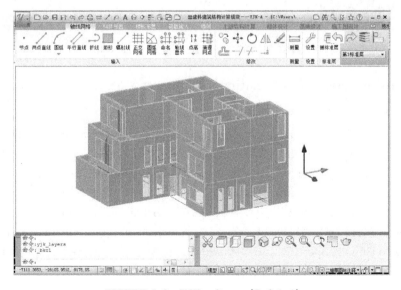

FIGURE 5.2 YJK software（Ref. 1. 4）

Guangsha structure software is another structural computing software for industrial and civil buildings developed by GSCAD Building Software in Shenzhen. Guangsha structure software was firstly introduced as integrated building structure software in 1996, which run in Windows operating system. The software mainly aims at the structural design of high-rise buildings. It can complete the modeling, calculation, automatic generation and processing of construction drawings. It can do the calculation of space thin arm linkage system SS and the calculation of space wall member system SSW. Guangsha steel structure software is divided into two major parts: the factory steel structure module (portal frame, plane truss and crane beam) and space structure software (Ref. 1.5). The software interface of Guangsha is shown in Figure 5.3.

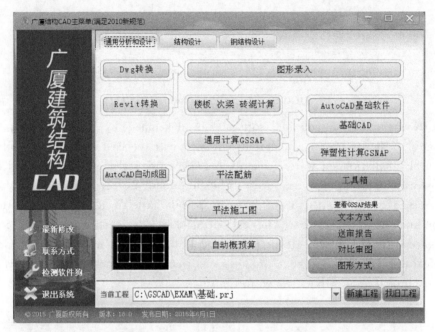

FIGURE 5.3　Guangsha software (Ref. 1.6)

5.1.2　Introduction of International Softwares

ETABS and SAP2000 are widely used for structural design beyond China.

The SAP2000 program is developed by the SAP (Structure Analysis Program) series initiated by the Edwards Wilson. The first generation products of SAP2000 were released in 1996. Then it was operated and updated by CSI Company. Users can use SAP2000 to perform all analysis and design work easily. The integrated design code can automatically generate wind loads, wave loads, bridge loads and seismic loads. It can automatically design and check complex steel structures and concrete structures according to Chinese specifications, American codes and other major international codes. The software interface of SAP2000 is shown in Figure 5.4 (Ref. 1.7).

Chapter 5　Application of Software

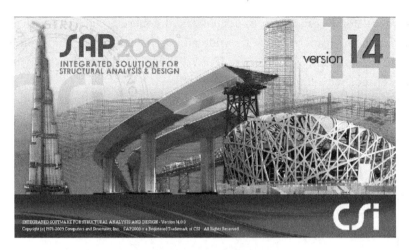

FIGURE 5.4　SAP2000 software

ETABS is also a structural design and analysis software which is widely used in the world. It is also developed by CSI Company. Based on 40 years development, the latest version of ETABS offers many technological advantages. It includes modeling and visualization tools for 3D objects, fast linear and nonlinear analysis capabilities, integrated design capabilities for a variety of materials and intuitive graphical display, reporting and construction drawing generation. Users can quickly and easily understand the analysis and design results. (Ref. 1.8) The software interface of ETABS is shown in Figure 5.5 (Ref. 1.9).

FIGURE 5.5　ETABS software

5.2　Example: Design of A Reinforced Concrete Structure

5.2.1　Introduction of Project

The example project is a six-storey building with rectangular plane. There are 4 grids in the X direction and each span is 22 inches long. There are 3 grids in the Y direction and each span is 18 feet long. Each storey is 12 inches high.

The lateral force resisting system consists of concrete frames and shear walls. The cross section of the beam is 24 inches wide and 30 inches deep. The column section is a square with a side length of 24 inches. The thickness of the wall is 14 inches. The floor is a concrete floor with 8 inches thick.

In addition to the self-weight of the structure, the additional dead load of the building is 25psf(1psf=4.8825kg/m^2), with live load of 80 psf for floors and roof. The Lateral loading is basically wind loads, which is based on the ASCE 7—10 code.

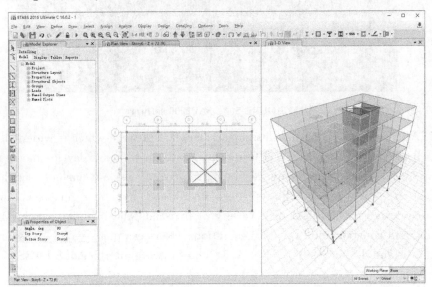

FIGURE 5.6 An example of a model

5.2.2 Begin a New Model

At this Step, we need to set dimensions and storey height. Then the vertical load and sectional dimensions of columns, beams and floors are defined.

STEP1. Start the program. The Start Page will display.

STEP2. Click the ***New Model*** button on the Start Page. The Model initialization dialogue box is shown in Figure 5.7.

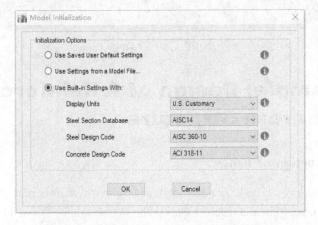

FIGURE 5.7 Model Initialization form

STEP3. Find the *Use Built-in Settings With*: option.

STEP4. Choose *U. S. Customary* from the Display Units drop down menu on the Model Initialization dialogue box. If you want to review the display units hold the mouse cursor over the *information* icon. If you want to change the initialization unit settings, click the *Options menu*>*Display Units* command.

STEP5. Choose *ACI 318 − 11* from the Concrete Design Code list. Click the *OK* button and the New Model Quick Templates dialogue box is displayed as shown in Figure 5.8.

The New Model Quick Templates dialogue is used to specify horizontal grid line spacing, storey data, and template models. The template provides a quick and simple way to initialize the model. They automatically add structural objects with appropriate attributes to the model. We highly recommend that you start your models using templates whenever possible. In this example, the model is built using the *Flat Slab with Perimeter Beams* template.

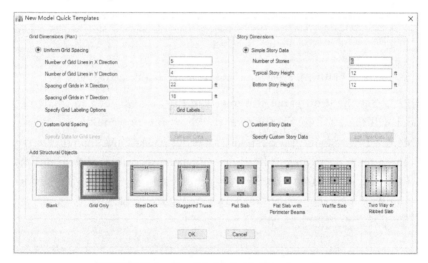

FIGURE 5.8 New Model Quick Templates form

STEP6. Set the number of grid lines in the Number of Grid Lines in X Direction edit box to **5**.

STEP7. Set the number of grid lines in the Number of Grid Lines in Y Direction edit box to **4**.

STEP8. Set the grid spacing in the Spacing of Grids in X Direction edit box to **22** ft (1ft=0.3048m).

STEP9. Set the grid spacing in the Spacing of Grids in Y Direction edit box to **18** ft.

STEP10. Set the number of storey in the Number of Storey edit box to **6**.

STEP11. Click the *Flat Slab with Perimeter Beams* button in the Add Structural Objects area to display the Structural Geometry and Properties for Flat Slab with Perimeter Beams dialogue box like Figure 5.9.

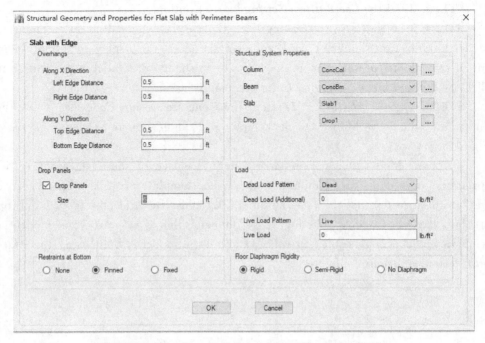

FIGURE 5.9 Structural Geometry and Properties form

STEP12. Set the drop panel size to **6** ft in the Size edit box in the Drop Panels area. This model will have 6-foot square drop panels at interior columns.

STEP13. In the Structural System Properties area, click the ellipsis button adjacent to the Column drop-down list. The Frame Properties form is shown in Figure 5.10.

STEP14. Click the **Add New Property** button in the Frame Properties form. The Frame Property Shape Type form is shown in Figure 5.11.

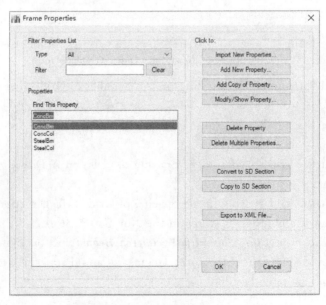

FIGURE 5.10 Frame Properties form

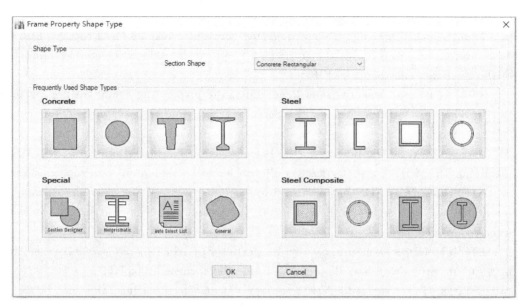

FIGURE 5.11　Frame Property Shape Type form

STEP15. Select the rectangle in the section shape drop down list of the shape type area, click the **OK** button, or click the rectangle section button under the concrete section in the common section type area. The Frame Section Property Data form is shown in Figure 5.12.

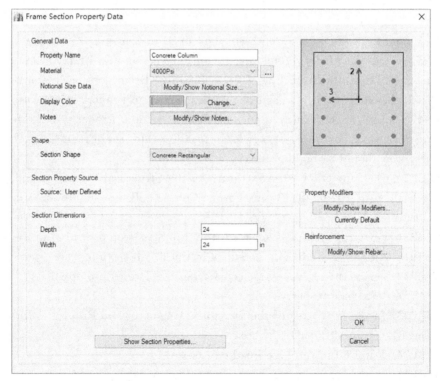

FIGURE 5.12　Frame Section Property Data form

15-1　Type **Concrete Column** in the Property Name edit box.

15-2 Set the value in the Depth edit box to **24** in.

15-3 Click the ***Modify/Show Modifiers*** button to display the Property/Stiffness Modification Factors form. If the section stiffness can be reduced to consider the effect of the cracking, and then click **OK**.

15-4 Click the ***Modify/Show Rebar*** button to display the Frame Section Property Reinforcement Data form. Review the default settings and then click the **OK** button to close the form.

15-5 Click the **OK** button on the Frame Section Property Data form to return to the Frame Properties form. Concrete Column should be shown on the Properties list.

STEP16. Click the **OK** button to return to the Structural Geometry and Properties for Flat Slab with Perimeter Beams form. Concrete Column should be shown as the selected Column section.

STEP17. In the Structural System Properties area, click the ***ellipsis*** button adjacent to the Beam drop-down list. The Frame Properties form will display.

STEP18. Click the ***Add New Property*** button in the Click area of the Frame Properties form. The Frame Property Shape Type form will appear.

STEP19. Select Rectangular from the Section Shape drop-down list in the Shape Type area and then click on the **OK** button, or click on the ***Rectangular Section*** button under Concrete in the Frequently Used Shape Types area of the Frame Property Shape Type form. The Frame Section Property Data form appears.

19-1 On the Frame Section Property Data from, type Concrete Beam in the Property Name edit box.

19-2 Set the value in the Depth edit box to 30 in (1in=0.0254m).

19-3 Click the ***Modify/Show Rebar*** button and on the Frame Section Property Reinforcement Data form select the M3 Design Only (Beam) option in the Design Type area. Click the **OK** button to return to the Frame Section Property Data form.

19-4 Click the **OK** button on the Frame Section Property Data form (after adjusting property modifiers, if needed) to return to the Frame Properties form. Concrete Beam should be highlighted.

STEP20. Click the **OK** button to return to the Structural Geometry and Properties for Flat Slab with Perimeter Beams form. Concrete Beam should be shown as the selected Beam section.

STEP21. Verify that Slab1 is selected from the Slab drop-down list in the Structural System Properties area. Slab1 is a default 8 in thick slab section.

STEP22. Verify that Drop1 is selected from the Drop drop-down list. Drop1 is a default 15 in thick slab property.

STEP23. In the Load area of the Structural Geometry and Properties form, type 25 in the Dead Load (Additional) edit box.

STEP24. Enter **80** in the Live Load edit box.

STEP25. In the rigid compartment of the floor, it is confirmed that the rigidity has been chosen. This option will ignore the relative deformation in the partition surface. Click the **OK** button to accept your selections and return to the New Model Quick

Templates form. At this point, with edge beams without beam floor button has a dark blue border highlights. Click the **OK** button on the New Model Quick Templates to display the model.

The model will appear in the two vertical paved view windows of the ETABS main window, and the left is the 3-D view on the right of the plane view, as shown in Figure 5.13. The number of view windows can be changed using the **Window List** button. View windows may be closed by clicking on the [**X**] button.

The Plan View is active in Figure 8, indicated by the highlighted display title tab. Set a view active by clicking anywhere in the view window. The location of the active Plan View is highlighted on the 3-D View by a Bounding Plane. The Bounding Plane may be toggled on/off using the **Options menu>Show Bounding Plane** command.

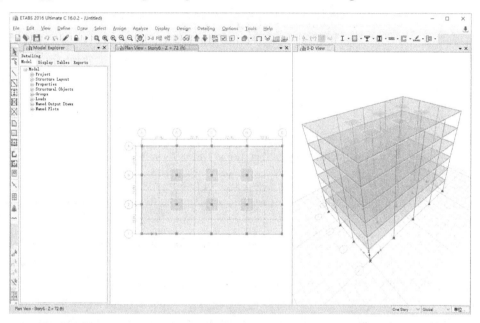

FIGURE 5.13 The ETABS main window with Plan View active

5.2.3 Add Floor Openings

In this Step, the program is set up to add openings to multiple storey simultaneously.

5.2.3.1 Set up to add objects to multiple storey simultaneously

Make sure that the plane view is active. To activate the window, move the cursor or mouse arrow to the view and click the left mouse button. When the view is active, the "display title bar" will be highlighted. Figure 5.13 shows the location of the display title bar.

STEP1. Click the Drawing & Selection drop-down list that reads "One Story" at the bottom right of the Main window, which is shown in Figure 8.

STEP2. Highlight *All Storey* in the list. This activates the All Storey option for drawing and selecting objects.

With the All Storey option active, as additions or changes are made to a storey, for

145

example storey 6, those additions and changes will apply to every storey in the building, Storey1 to storey 6.

5.2.3.2 Draw shell objects

Make sure that the plan view is active.

STEP1. Click the **Draw Rectangular Floor/Wall** button or use the **Draw menu > Draw Floor/Wall Objects > Draw Rectangular Floor** command. The Properties of Object form for shells as shown in Figure 5.14 will display "docked" in the lower left-hand corner of the program.

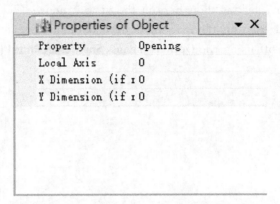

FIGURE 5.14 Properties of Object form for shells

STEP2. Click once in the drop-down list opposite the Property item to activate it and then select Opening in the resulting list. Opening will create a void where the shell is drawn.

STEP3. Check that the **Snap to Grid Intersections and Points** command is active. This will assist in accurately drawing the shell object. Alternatively, use the **Draw menu > Snap Options** command to ensure that these snaps are active. By default, this command is active.

STEP4. In the plan view, click once at column C-3, and while holding down the *left mouse* button, drag the cursor to column D-2. Release the mouse button to draw a rectangular opening.

If you have made a mistake while drawing this object, click the Select Object button, to change the program from Draw mode to Select mode. Then click the **Edit menu > Undo Shell Object Add** command and repeat Action Items A thru D to re-draw the openings.

STEP5. Click the **Select Object** button to change the program from Draw mode to Select mode.

STEP6. Hold down the Ctrl key on your keyboard and left click once in the Plan View on column C-3. A selection list similar to the one shown in Figure 5.15 pops up because multiple objects exist at the location that was clicked. In this example, a joint object, a column object, an opening and two floor objects (drop and slab) exist at the same location. Note that the selection list will only appear when the Ctrl key is used with the left click.

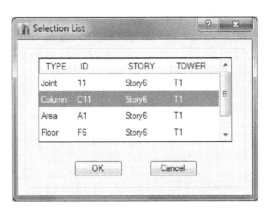

FIGURE 5.15　Selection List form

STEP7. Select the column from the list by clicking on it and then on the **OK** button. The column at C-3 is now selected, as indicated by the dashed lines in the 3-D view. It is selected over its entire height because the All Storey feature is active. Note that the status bar in the bottom left-hand corner of the main ETABS window indicates that 6 frames have been selected.

STEP8. Repeat the column selection process at D-3, D-2, and C-2. The status bar should indicate that 24 frames have been selected.

STEP9. Press the Delete key on your keyboard or click the Edit menu＞Delete command to delete the selection because no columns should exist at these four locations.

The model should now appear as shown in Figure 5.16.

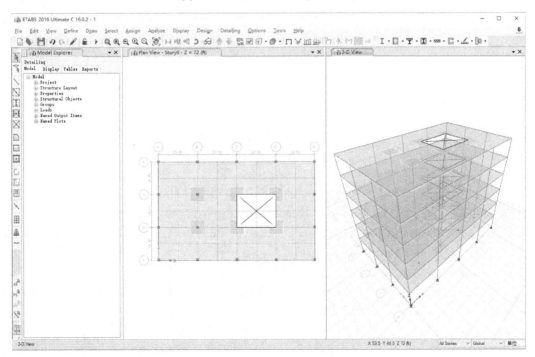

FIGURE 5.16　The example model with the slab openings drawn

5.2.3.3 Save the model

Click the ***File menu*** > ***Save*** command, or the ***Save*** button, to save your model. Specify the directory in which you want to save the model, for this example specify the file name as Concrete Building.

5.2.4 Add Walls

In this Step, a wall stack is added to model the interior walls.

STEP1. Click the ***Draw menu*** > ***Draw Wall Stacks*** command, or the ***Draw Wall Stacks*** button, to access the New Wall Stack form shown in Figure 5.17.

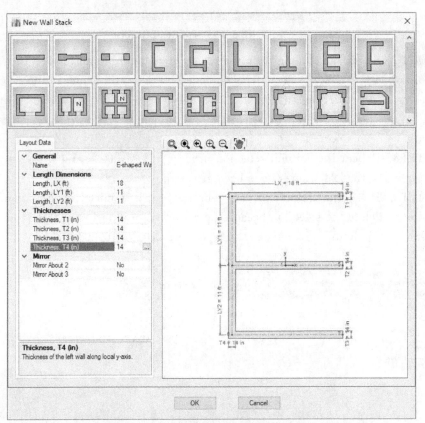

FIGURE 5.17 New Wall Stack form

STEP2. Click on the ***E-shaped Wall*** button to display a wall arrangement.

STEP3. On the Layout Data tab, click in the Length, LX (ft) edit box and type **18**.

STEP4. Type **11** into the Length, LY1 (ft) edit box.

STEP5. Type **11** into the Length, LY2 (ft) edit box.

STEP6. Type **14** into all four of the Thicknesses edit boxes—all of the walls should be 14 inches thick.

STEP7. Click the ***OK*** button. The Properties of Object form for Wall Stacks shown in Figure 5.18 will display "docked" in the lower left-hand corner of the main window.

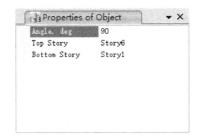

FIGURE 5.18 Properties of Object form for Wall Stack objects

STEP8. Click in the Angle edit box on the Properties of Object form, set the angle to **90**, and press the Enter key on your keyboard. This will rotate the wall stack object 90 degrees from the default position.

STEP9. Left click once in the Plan View such that the top-left corner of the wall stack shown using dashed lines is located at C-3. The wall stack should match the geometry of the slab opening.

Notice that the wall stack spans the entire height of the building as the Top Storey and Bottom Story drop-down lists in the Properties of Object form were set to Storey 6 and Storey 1, respectively.

STEP10. Click the Select Object button, change the program from Draw mode to select mode.

STEP11. Click the *File menu* > *Save* command to save your model. The model is shown in Figure 5.19.

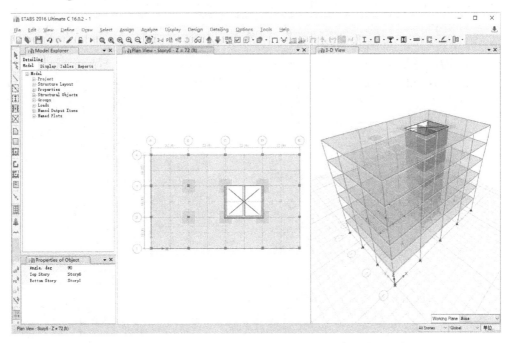

FIGURE 5.19 The example model with the wall stack drawn

Design of Concrete Structure and Software Application

5.2.5 Define Static Load Patterns

The static loads used in this example consist of the dead, live, and wind loads. The number of load patterns can be defined.

At the beginning of the tutorial, the dead load consists of self-weight of the building plus an additional 25 psf. The live load of 80 psf was also assigned. The ASCE 7−10 wind load that will be applied to the building and will be automatically calculated by the program.

STEP1. In the Model Explorer window, click on the **Loads** node on the Model tab to expand the tree. If the Model Explorer is not displayed, click the **Options menu** > **Show Model Explorer** command.

STEP2. On the expanded tree, right-click on the **Load Patterns** branch to display a context sensitive menu. On this menu, click on the **Add New Load Pattern** command to display the Define Load Patterns form. Note that the Dead and Live load patterns are already defined.

STEP3. Click in the Load edit box and type the name of the new load pattern, **Wind**.

STEP4. Select Wind from the Type drop-down list. Make sure that the Self Weight Multiplier is set to zero for the Wind load pattern. Self-weight should typically be included in only one load pattern.

STEP5. Use the Auto Lateral Load drop-down list to select ASCE 7−10; with this option ETABS will automatically apply wind load based on the ASCE 7 − 10 code requirements.

STEP6. Click the Add New Load button.

STEP7. With the Wind load pattern highlighted, click the Modify Lateral Load button. This will display the ASCE 7−10 Wind Load Pattern form as shown in Figure 5.20.

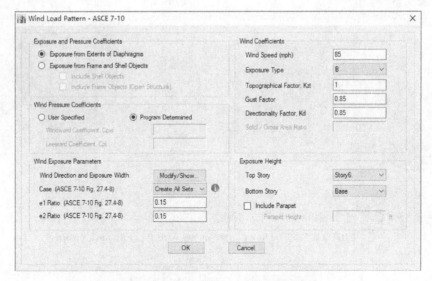

FIGURE 5.20 Wind Load Pattern form

7-1. Select the ***Exposure from Extents of Rigid Diaphragms option*** in the Exposure and Pressure Coefficients area.

The Exposure from Extents of Rigid Diaphragms option means that the program will automatically calculate all the wind load cases as prescribed in the ASCE 7 — 10 code. In this example, there will be a total of 12 different load permutations of varying magnitude and direction that will be applied to the rigid floor diaphragms. Hold the mouse cursor over the information icon to display a table of the different wind cases.

7-2. Verify that the Case drop-down list in the Wind Exposure Parameters area shows Create All Sets.

7-3. Review all the other wind parameters, including the exposure heights, and then click the OK button to return to the Define Load Patterns form displayed in Figure 5.21.

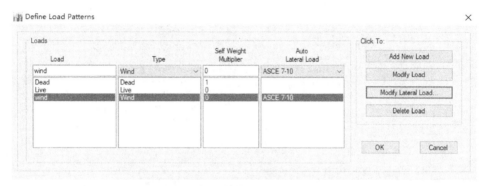

FIGURE 5.21   Define Load Patterns form

STEP8. Click the ***OK*** button to accept the load patterns.

STEP9. Click the File menu＞Save command, or the Save button, to save the model.

5.2.6   Review Diaphragms

In this Step, the extent of the rigid floor diaphragms will be displayed. Rigid diaphragms are typically used to model floor systems that have a large stiffness in-plane by removing the in-plane degrees of freedom. A rigid diaphragm has no in-plane deformations, and therefore, no in-plane shell stresses are reported by the program. However, in reality, these diaphragms do carry in-plane forces (see the ***Display menu＞Force/Stress Diagrams＞Diaphragm Forces*** command), and thus users should make sure that they design and detail the diaphragms for these forces, e.g. by using chords and collectors to transfer forces from the diaphragms into the lateral resisting frames and walls.

Make sure that the Plan View is active.

STEP1. Click the ***Set Display Options*** button, or use the ***View menu＞Set Display Options*** command. The Set View Options form is shown in Figure 5.22.

FIGURE 5.22 Set View Options form

STEP2. On the General tab, check the Diaphragm Extent option in the Other Special Items area.

STEP3. Click the **OK** button to display the rigid diaphragm links as illustrated by the dashed lines radiating out from the diaphragm centre.

STEP4. Right-click anywhere on the slab (but not on a beam, drop, or wall) to display the Slab Information form as shown in Figure 5.23.

Chapter 5　Application of Software

FIGURE 5.23　Slab Information form

STEP5. Click on the Assignments tab on the Slab Information form and note that the Diaphragm assignment is D1.

STEP6. Click the **OK** button to close the form.

STEP7. Click the ***Define menu*** > ***Diaphragms*** command to display the Define Diaphragm form.

7-1. On the Define Diaphragm form, highlight D1 in the Diaphragms area and click the Modify/Show Diaphragm button. The Diaphragm Data form is shown in Figure 5.24.

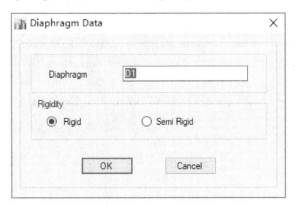

FIGURE 5.24　Diaphragm Data form Step

7-2. Verify that the Rigid option is selected in the Rigidity area to ensure that the floor slabs of this building will not have any deformations in plane.

7-3. Click the **OK** button to return to the Define Diaphragm form.

STEP8. Click the **OK** button to close the Define Diaphragm form.

STEP9. Click the **Set Display Options** button or use the **View menu > Set Display Options** command and the Set View Options form will appear.

STEP10. Uncheck the Diaphragm Extent option.

STEP11. Click the **OK** button to close the Set View Options form.

STEP12. Click the **File menu > Save** command, or the **Save** button, to save the model.

5.2.7 Review the Load Cases

In this Step, the load cases generated from the load patterns will be reviewed.

STEP1. Click the **Define menu > Load Cases** command to display the Load Cases form as shown in Figure 5.25.

Notice that three load cases are listed, one for each of the three load patterns that were defined: Dead, Live and Wind.

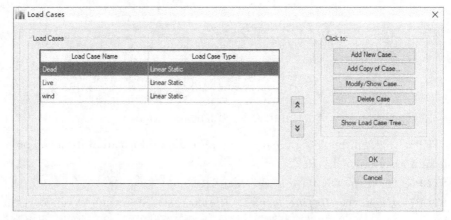

FIGURE 5.25 Load Cases form

STEP2. On the Load Cases form, highlight Wind in the Load Cases area and click the **Modify/Show** Case button. The Load Case Data form shown in Figure 5.26 will display.

2-1. Verify that Linear Static is selected in the Load Case Type dropdown list.

2-2. Verify that Wind is shown as the Load Pattern in the Loads Applied area.

2-3. Select the Use Preset P-Delta Settings option in the P-Delta/Nonlinear Stiffness area and click the **Modify/Show** button to display the Preset P-Delta Options form.

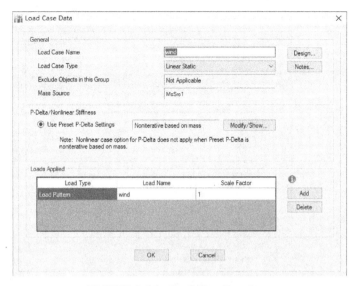

FIGURE 5.26 Load Case Data form

2-4. On the Preset P-Delta Options form, select the Non-iterative—Based on Mass option in the Automation Method area. This will add P-Delta effects to the analysis. Click the **OK** button to close the Preset P-Delta Options form and return to the Load Case Data form.

2-5. The Use Preset P-Delta Settings value should now show Noniterative based on mass. Click the **OK** button on the Load Case Data form.

STEP3. Click the **OK** button to close the Load Cases form

STEP4. Click the **File menu > Save** command, or the Save button, to save the model.

5.2.8 Run the Analysis

In this Step, the analysis will be run.

STEP1. Click the **Analyze menu > Check Model** command. The Check Model form shown in Figure 5.27 will display.

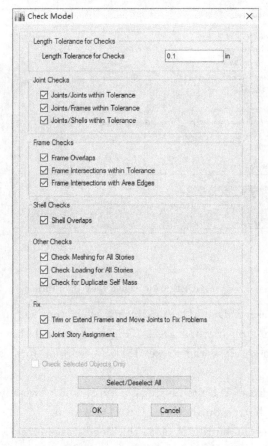

FIGURE 5.27 Check Model form

STEP2. Check all check boxes and click the **OK** button. A warning message similar to that shown in Figure 5.28 should display indicating that the model has no connectivity issues.

FIGURE 5.28 Check Model Warning

STEP3. Click the ***Analyze menu*** > ***Run Analysis*** command or the Run Analysis button .

The program will create the analysis model from your object-based input. After the analysis has been completed, the program will automatically display a deformed shape similar to that shown in Figure 5.29, and the model is locked. The model is locked when the ***Lock/Unlock Model*** button, appears locked. Locking the model prevents any changes to the model that would invalidate the analysis results.

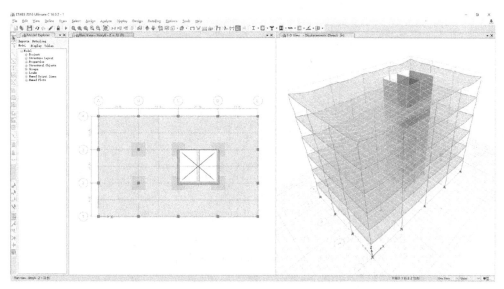

FIGURE 5.29   Deformed shape display

5.2.9 Display the Results

In this Step, the analysis results will be displayed and reviewed.

STEP1. Make sure that the 3D View is active—this can be done by clicking on the Display Title Tab.

STEP2. Click on the ***Display Shell Stresses/Forces...*** button, or the ***Display menu*** > ***Force/Stress Diagrams*** > ***Shell Stresses/Forces...*** command to access the Shell Forces/Stresses form shown in Figure 5.30.

FIGURE 5.30   Shell Forces/Stresses form

2-1. Select *Dead* from the Load Case drop-down list.

2-2. Select *Resultant Forces* as the Component Type.

2-3. Select the *M11* component.

2-4. Select *Display on Deformed Shape* from the Contour Option drop-down list.

2-5. Check the *Show Fill* checkbox.

2-6. Click the *OK* button to generate the moment contours shown in Figure 5.31.

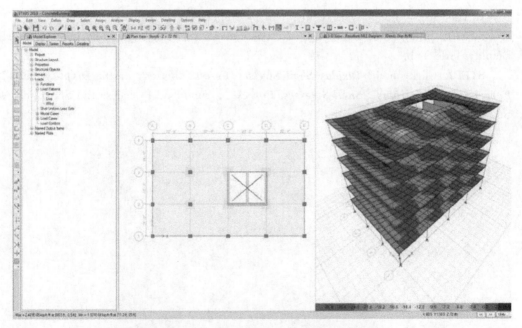

FIGURE 5.31   M11 moment contours in a 3D view

STEP3. Make the Plan View active by clicking on the Plan View Title Tab.

STEP4. Click the ***Show Deformed Shape*** button, or the ***Display menu>Deformed Shape*** command to display the Deformed Shape form.

4-1. On the Deformed Shape form, select Wind from the Load Case drop-down list.

4-2. Set the Step Number to 1—there should be a total of 12 steps available for the Wind load case.

4-3. Click the ***OK*** button to display the deformed shape shown in Figure 5.32.

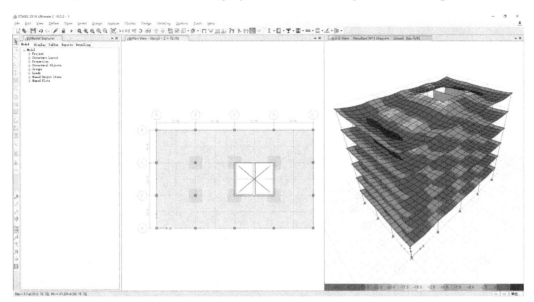

FIGURE 5.32 Deformed Shape for Wind load case

Note that even though the wind load for Step 1 (Set 1) is applied in the X direction, there is a rotation of the structure due to the lateral stiffness eccentricity caused by the non-symmetric layout of the walls.

STEP5. Each of the deformed shapes due to the 12 different wind load permutations (as specified by the ASCE 7−10 code) may be viewed by clicking on the VCR buttons, located in the lower right-hand corner of the display.

STEP6. After reviewing the different deformed shapes, set the plan view back to an undeformed view by clicking on the ***Show Undeformed Shape*** button, or the ***Display menu>Undeformed Shape*** command.

5.2.10 Design the Concrete Frames

In this Step, the concrete beams and columns will be designed. Note that the analysis (Step 7) should be run before performing design.

STEP1. Make sure that the Plan View is active—this can be done by clicking on the Display Title Tab.

STEP2. In the Plan View, right click on the perimeter beam along grid line 1 between grids A and B. The Beam Information form is shown in Figure 5.33. Note that the Design tab reports that the Design Procedure is Concrete Frame Design.

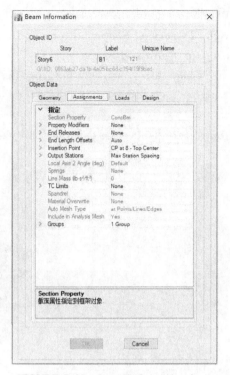

FIGURE 5.33 Beam Information form

STEP3. Click the ***Cancel*** button to close the Beam Information form.

STEP4. Click the ***Design menu***＞***Concrete Frame Design***＞***View/Revise Preferences*** command. The Concrete Frame Design Preferences form is shown in Figure 5.34 displays.

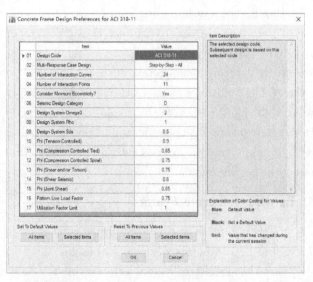

FIGURE 5.34 Concrete Frame Design Preferences form

4-1. Verify that the Design Code drop-down list is set to ACI 318−11.

4-2. Review the design parameters shown on this form and then click the ***OK*** button to accept any changes made.

STEP5. With the Plan View active, click the ***Design menu* > *Concrete Frame Design* > *Start Design/Check*** command to start the design process. The program designs the concrete beams and columns, specifying the required reinforcing based on the shape and size of the members defined in Step 1. When the design is complete, the longitudinal reinforcing is displayed on the model. The model is shown in Figure 5.35.

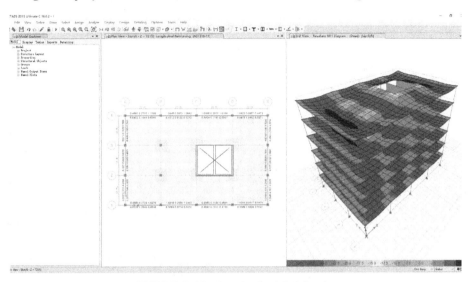

FIGURE 5.35 Longitudinal Reinforcing

STEP6. With the Plan View active, click on the Move Down in List button until Plan View—Story 1 is displayed.

STEP7. Right click on one of the perimeter beams in the Plan View shown in Figure 5.35. The Concrete Beam Design Information form is shown in Figure 5.36.

This form shows the required reinforcing steel calculated for each design load combination at different locations along the length of the beam. The largest reinforcing value, either top or bottom steel, is highlighted.

Concrete Beam Design Information (ACI 318-11)

Story: Story1 Section Name: ConcBm
Beam: B11

COMBO ID	STATION LOC	TOP STEEL	BOTTOM STEEL	SHEAR STEEL
DCon1	2.3750	0.2325	0.1436	0.0000
DCon1	4.0000	0.1436	0.1436	0.0000
DCon1	4.0000	0.1436	0.1436	0.0000
DCon1	5.8125	0.1436	0.2825	0.0000
DCon1	7.6250	0.1436	0.4375	0.0000
DCon1	7.6250	0.1436	0.4421	0.0000
DCon1	9.4375	0.1436	0.4577	0.0000
DCon1	11.2500	0.1436	0.4443	0.0000
DCon1	11.2500	0.1436	0.4408	0.0318
DCon1	13.0625	0.1436	0.2987	0.0318
DCon1	14.8750	0.1436	0.1436	0.0318
DCon1	14.8750	0.1436	0.1436	0.0000
DCon1	16.0625	0.1436	0.1436	0.0000
DCon1	17.2500	0.2356	0.1176	0.0000
DCon2	0.7500	0.7921	0.3934	0.0000

Overwrites Summary Flex. Details Shear Envelope

OK Cancel

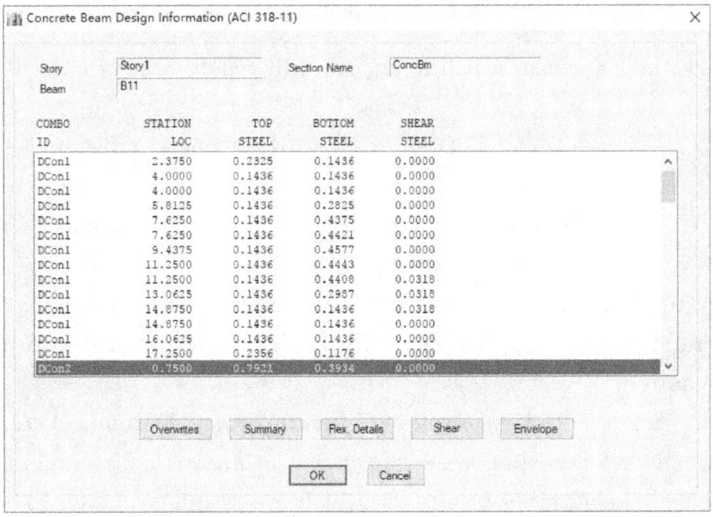

FIGURE 5.36 Concrete Beam Design Information form

7-1. Click the Envelope button on the Concrete Beam Design Information form. The Beam Section Design Report is shown in Figure 5.37. This report shows detailed design information about the beam. Review the information on each page. Note that you can print this information using the **Print Report** button, on the menu bar.

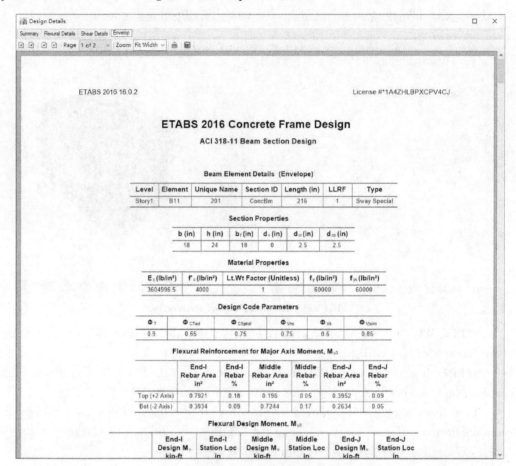

FIGURE 5.37　Beam Section Design Report

7-2. Click the [**X**] button in the upper right-hand corner of the Report Viewer to close it.

STEP8. Click the **Cancel** button to close the Concrete Beam Design Information form.

STEP9. Click the **Set Elevation View** button, and select A and click the **OK** button to set the view to an Elevation View of grid line A.

The elevation view shows the area of the longitudinal reinforcing required for the columns. A right click on any column will display more detailed information about the reinforcing required.

STEP10. Click the **Design menu > Concrete Frame Design > Verify All Members Passed** command. A form similar to that shown in Figure 5.38 should appear indicating that the shape and reinforcing of all concrete frame members is adequate. Click the **OK** button to close the form.

FIGURE 5.38 Verify All Members Passed message

STEP11. Click the ***Assign menu*** > ***Clear Display of Assigns*** command to clear the longitudinal reinforcing display.

STEP12. Make the 3D View active by clicking on the Display Title Tab.

STEP13. Click the ***Display menu*** > ***Undeformed Shape*** command or the **Show Undeformed Shape** button, to clear the display of the moment diagram.

STEP14. Click the ***File menu*** > ***Save*** command, or the **Save** button, to save your model.

5.2.11 Design the Shear Walls

In this Step, the concrete shear walls will be designed. Note that the analysis (Step 7) should be run before performing design.

STEP1. Make the Elevation View active by clicking on the Elevation View Title Tab.

STEP2. Click the ***Set Plan View*** button, select Story6 and click the **OK** button to set the view to a Plan View of Story6.

STEP3. Click the ***Select menu*** > ***Select*** > ***Object Type*** command and the Select by Object Type form will display.

3-1. On the Select by Object Type form, highlight Walls.

3-2. Click the ***Select*** button, and then the ***Close*** button.

STEP4. Click the ***View menu*** > ***Show Selected Objects Only*** command to display only the elevator core walls in the Plan View.

STEP5. Click the ***Define menu*** > ***Pier Labels*** command. The Pier Labels form shown in Figure 5.39 appears.

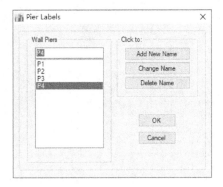

FIGURE 5.39 Pier Labels form

5-1. Type **P3** in the Wall Piers edit box and click the **Add New Name** button.

5-2. Type **P4** in the Wall Piers edit box and click the **Add New Name** button.

Note that pier labels P1 and P2 are predefined by the program, with P2 pre-assigned to all walls in the wall stack.

5-3. Click the **OK** button to close the form.

STEP6. Make sure that All Storey is still selected in the Drawing & Selection drop-down list in the lower right-hand corner of the Main window.

STEP7. In the Plan View, click on the wall that lies along grid line C. The status bar should indicate that 6 shells have been selected.

STEP8. Click the **Assign menu > Shell > Pier Label** command and the Shell Assignment—Pier Label form will display.

8-1. On the Shell Assignment—Pier Label form, highlight P1 and click the **Apply** button.

8-2. Without closing the Shell Assignment—Pier Label form, click on the middle wall parallel to grid line C that is set between grid lines C and D. Highlight P2 on the Shell Assignment—Pier Label form and click the **Apply** button.

8-3. Without closing the Shell Assignment—Pier Label form, click on the wall that lies along grid line D. Highlight P3 in the Piers area and click the **Apply** button.

8-4. Without closing the Shell Assignment—Pier Label form, "window" around the wall that lies on grid line 2. To "window", click the left mouse button above and to the left of grid intersection C2 and then, while holding the left mouse button down, drag the mouse until it is below and to the right of grid intersection D-2. A selection box similar to that shown in Figure 5.40 should expand around the wall as the mouse is dragged across the model. Release the left mouse button and the program will select the wall— the status bar should indicate that 12 shells have been selected.

8-5. Highlight P4 on the Shell Assignment—Pier Label form and click the **Apply** button.

8-6. Click the **Close** button to close the Shell Assignment—Pier Label form.

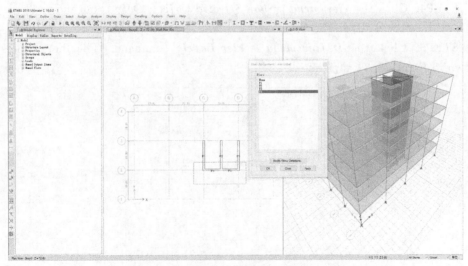

FIGURE 5.40 Selecting Walls

STEP9. With the Plan View active, click the **Set Default 3D View** button to show the core in a 3-D view.

STEP10. Click the **Design menu > Shear Wall Design > View/Revise Preferences** command. The Shear Wall Design Preferences form will display.

10-1. Verify that the Design Code drop-down list is set to ACI 318—11.

10-2. Review the design parameters shown on this form and then click the **OK** button to accept any changes made.

STEP11. Click the **Design menu > Shear Wall Design > Start Design/Check** command to start the design process. The program designs the shear walls, specifying the required reinforcing based on the shape and size of the members defined in Step 3.

When the design is complete, the pier longitudinal reinforcing is displayed on the model. The model appears as shown in Figure 5.41.

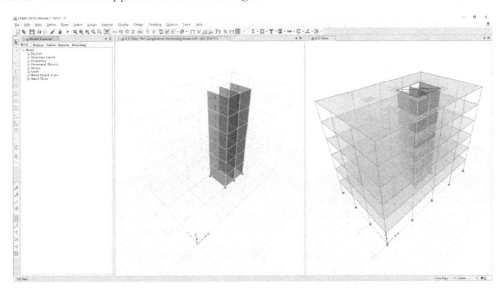

FIGURE 5.41 Pier Longitudinal Reinforcing

STEP12. Zoom in on the walls using the **Rubber Band Zoom** button.

STEP13. Right click on one of the walls in the Pier Longitudinal Reinforcing Areas view to display the Shear Wall Design Report.

This report shows detailed design information about the pier. Click the [**X**] button in the top right corner to close the report.

STEP14. With the Pier Longitudinal Reinforcing Areas view active, click the **Assign menu > Clear Display of Assigns** command.

STEP15. Click the **Set Plan View** button, select **Story 6** and click the **OK** button.

STEP16. Click the **View menu > Show All Objects** command.

STEP17. Make sure that the Model Explorer window is visible; if not, click the **Options menu > Show Model Explorer** command.

STEP18. Click the Tables tab in the Model Explorer to display the table tree. Click on the **Tables** node and then on the **Design** node to expand the tree.

STEP19. Click on the **Shear Wall Design** node to expose the Shear Wall Pier

Summary leaf. Right click on the Shear Wall Pier Summary leaf and from the context sensitive menu select ***Show Table***. A table summarizing the design of the shear wall piers now appears across the bottom of the main window as shown in Figure 5.42.

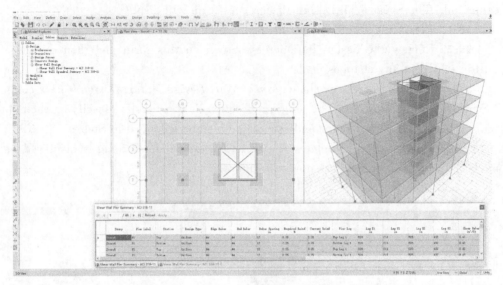

FIGURE 5.42　Shear Wall Pier Summary table

STEP20. Click the [***X***] button on the title bar of the Shear Wall Pier Summary table to close the table.

STEP21. Click the ***File menu*** > ***Save*** command, or the ***Save*** button, to save your model. This tutorial is now complete.

References

5.1　Li Bo. PKPM building structure design from introductory to mastery[M]. Beijing: Machinery Industry Press, 2015.

5.2　https://www.pkpm.cn/.

5.3　Lin Bai. Application and example analysis of YJK[M]. Beijing: China Construction Industry Press, 2018.

5.4　http://www.yjk.cn/.

5.5　Tan Yiping. General purpose analysis and design software course for building structure[M]. 3th ed. Beijing: China Construction Industry Press, 2016.

5.6　http://www.gscad.com.cn/.

5.7　http://www.cisec.cn/SAP2000/SAP2000.aspx.

5.8　Yang Yong. Detailed explanation of ETABS structure design example[M]. Beijing: China Construction Industry Press, 2015.

5.9　http://www.cisec.cn/ETABS/ETABS.aspx.

Appendix A

TABLE A.1   Reference table of wind pressure coefficient in some cities

Provincial, Municipal names	City, Town names	Height above sea level/m	Wind pressure/(kN/m²)		
			$n=10$	$n=50$	$n=100$
Jiangsu	Najing	8.9	0.25	0.4	0.45
	Xuzhou	41	0.25	0.35	0.4
	Ganyu	2.1	0.3	0.45	0.5
	Huaiyin	17.5	0.25	0.4	0.45
	Sheyang	2	0.3	0.4	0.45
	Zhenjiang	26.5	0.3	0.4	0.45
	Wuxi	6.7	0.3	0.45	0.5
	Taizhou	6.6	0.25	0.4	0.45
	Lianyungang	3.7	0.35	0.55	0.65
	Yancheng	3.6	0.25	0.45	0.55
	Gaoyou	5.4	0.25	0.4	0.45
	Dongtai	4.3	0.3	0.4	0.45
	Lvsi, Qidong	5.5	0.35	0.5	0.55
	Changzhou	4.9	0.25	0.4	0.5
	Liyang	7.2	0.25	0.4	0.5
	Dongshan, Wuzhong	17.5	0.3	0.45	0.5
Zhejiang	Hangzhou	41.7	0.3	0.45	0.5
	Tianmushan, Lin'an	1505.9	0.55	0.7	0.8
	Zhapu, Pinghu	5.4	0.35	0.45	0.5

Continued

Provincial, Municipal names	City, Town names	Height above sea level/m	Wind pressure/(kN/m²)		
			$n=10$	$n=50$	$n=100$
Zhejiang	Cixi	7.1	0.3	0.45	0.5
	Shengxi	79.6	0.85	1.3	1.55
	Shengshan, Shengxi	124.6	0.95	1.5	1.75
	Zhoushan	35.7	0.5	0.85	1
	Jinhua	62.6	0.25	0.35	0.4
	Shengxian	104.3	0.25	0.4	0.5
	Ningbo	4.2	0.3	0.5	0.6
	Shipu, Xiangshan	128.4	0.75	1.2	1.4
	Quzhou	66.9	0.25	0.35	0.4
	Lishui	60.8	0.2	0.3	0.35
	Longquan	198.4	0.2	0.3	0.35
	Kuocangshan, Linhai	1383.1	0.6	0.9	1.05
	Wenzhou	6	0.35	0.6	0.7
	Hongjia, Jiaojiang	1.3	0.35	0.55	0.65
	Kanmen, Yuhuan	95.9	0.7	1.2	1.45
	Beiji, Rui'an	42.3	0.95	1.6	1.9
Guangdong	Guangzhou	6.6	0.3	0.5	0.6
	Nanxiong	133.8	0.2	0.3	0.35
	Lianxian	97.6	0.2	0.3	0.35
	Shaoguan	69.3	0.2	0.35	0.45
	Fogang	67.8	0.2	0.3	0.35
	Lianping	214.5	0.2	0.3	0.35
	Meixian	87.8	0.2	0.3	0.35
	Guangning	56.8	0.2	0.3	0.35
	Goyao	7.1	0.3	0.5	0.6
	Heyuan	40.6	0.2	0.3	0.35
	Huiyang	22.4	0.35	0.55	0.6
	Wuhua	120.9	0.2	0.3	0.35
	Shantou	1.1	0.5	0.8	0.95
	Huilai	12.9	0.45	0.75	0.9
	Nan'ao	7.2	0.5	0.8	0.95

Continued

Provincial, Municipal names	City, Town names	Height above sea level/m	Wind pressure/(kN/m²)		
			$n=10$	$n=50$	$n=100$
Guangdong	Xinyi	84.6	0.35	0.6	0.7
	Luoding	53.3	0.2	0.3	0.35
	Taishan	32.7	0.35	0.55	0.65
	Shenzhen	18.2	0.45	0.75	0.9
	Shanwei	4.6	0.5	0.85	1
	Zhanjian	25.3	0.5	0.8	0.95
	Yangjiang	25.3	0.5	0.7	0.8
	Dianhai	11.8	0.45	0.7	0.8
	Shanchuandao, Taishan	21.5	0.75	1.05	1.2
	Xuwen	67.9	0.45	0.75	0.9

Appendix B

TABLE B.1 Stirrup's maximum spacing in a beam s_{max} mm

Beam's height h	$150<h\leqslant300$	$300<h\leqslant500$	$500<h\leqslant800$	$h>800$
$V\leqslant0.7f_tbh_0$	200	300	350	400
$V>0.7f_tbh_0$	150	200	250	300

TABLE B.2 Stirrup's Minimum diameter in the beam mm

Beam's height	$h\leqslant800$	$h>800$
Stirrup's diameter	6	8

Two-way slab's calculation coefficient

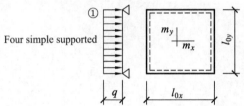

Four simple supported

B-1

l_{0x}/l_{0y}	f	m_x	m_y	l_{0x}/l_{0y}	f	m_x	m_y
0.50	0.01013	0.0965	0.0174	0.80	0.00606	0.0561	0.0334
0.55	0.00940	0.0892	0.0210	0.85	0.00547	0.0506	0.0348
0.60	0.00867	0.0820	0.0242	0.90	0.00496	0.0456	0.0358
0.65	0.00796	0.0750	0.0271	0.95	0.00449	0.0410	0.0364
0.70	0.00727	0.0683	0.0296	1.00	0.00406	0.0368	0.0368
0.75	0.00663	0.0620	000317				

Appendix

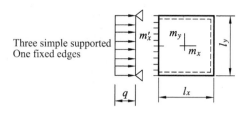

Three simple supported One fixed edges

B-2

l_x/l_y	l_y/l_x	f	f_{max}	m_x	m_{xmax}	m_y	m_{ymax}	m'_x
0.50		0.00488	0.00504	0.0583	0.0646	0.0060	0.0063	−0.1212
0.55		0.00471	0.00492	0.0563	0.0618	0.0081	0.0087	−0.1187
0.60		0.00453	0.00472	0.0539	0.0589	0.0104	0.0111	−0.1158
0.65		0.00432	0.00448	0.0513	0.0559	0.0126	0.0133	−0.1124
0.70		0.00410	0.00422	0.0485	0.0529	0.0148	0.0154	−0.1087
0.75		0.00388	0.00399	0.0457	0.0496	0.0168	0.0174	−0.1048
0.80		0.00365	0.00376	0.0428	0.0463	0.0187	0.0193	−0.1007
0.85		0.00343	0.00352	0.0400	0.0431	0.0204	0.0211	−0.0965
0.90		0.00321	0.00329	0.0372	0.0400	0.0219	0.0226	−0.0922
0.95		0.00299	0.00306	0.0345	0.0369	0.0232	0.0239	−0.0880
1.00	1.00	0.00279	0.00285	0.0319	0.0340	0.0243	0.0249	−0.0839
	0.95	0.00316	0.00324	0.0324	0.0345	0.0280	0.0287	−0.0882
	0.90	0.00360	0.00368	0.0328	0.0347	0.0322	0.0330	−0.0926
	0.85	0.00409	0.00471	0.0329	0.0347	0.0370	0.0378	−0.0970
	0.80	0.00464	0.00473	0.0326	0.0343	0.0424	0.0433	−0.1014
	0.75	0.00526	0.00536	0.0319	0.0335	0.0485	0.0494	−0.1056
	0.70	0.00595	0.00605	0.0308	0.0323	0.0553	0.0562	−0.1096
	0.65	0.00670	0.00680	0.0291	0.0306	0.0627	0.0637	−0.1133
	0.60	0.00752	0.00762	0.0268	0.0289	0.0707	0.0717	−0.1166
	0.55	0.00838	0.00848	0.0239	0.0271	0.0792	0.0801	−0.1193
	0.50	0.00927	0.00935	0.0205	0.0249	0.0880	0.0888	−0.1215

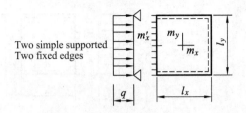

B-3

l_x/l_y	l_y/l_x	f	m_x	m_y	m_x'
0.50		0.00261	0.0416	0.0017	−0.0943
0.55		0.00259	0.0410	0.0028	−0.0840
0.60		0.00255	0.0402	0.0042	−0.0834
0.65		0.00250	0.0392	0.0057	−0.0826
0.70		0.00243	0.0379	0.0072	−0.0814
0.75		0.00236	0.0366	0.0088	−0.0799
0.80		0.00228	0.0351	0.0103	−0.0782
0.85		0.00220	0.0335	0.0118	−0.0763
0.90		0.00211	0.0319	0.0133	−0.0743
0.95		0.00201	0.0302	0.0146	−0.0721
1.00	1.00	0.00192	0.0285	0.0158	−0.0698
	0.95	0.00223	0.0296	0.0189	−0.0746
	0.90	0.00260	0.0806	0.0224	−0.0797
	0.85	0.00303	0.0314	0.0266	−0.0850
	0.80	0.00354	0.0319	0.0316	−0.0904
	0.75	0.00413	0.0321	0.0374	−0.0959
	0.70	0.00482	0.0318	0.0441	−0.1013
	0.65	0.00560	0.0308	0.0518	−0.1066
	0.60	0.00647	0.0292	0.0604	−0.1114
	0.55	0.00743	0.0267	0.0698	−0.1156
	0.50	0.00844	0.0234	0.0798	−0.1191

Appendix C

TABLE C.1 Maximum spacing between expansion joints of reinforced concrete structure

Structure type		Indoors or in soil	Open-air
Bent structure	prefabricated	100	70
Frame structure	prefabricated	75	50
	cast-in-place type	55	35
Shear wall structure	prefabricated	65	40
	cast-in-place type	45	30
The structure of the retaining wall, the basement wall, and so on	prefabricated	40	30
	cast-in-place type	30	20

Note:

① In the case of sufficient basis and reliable measures, the value of the table can be increased and reduced.

② The distance between the expansion joints of the assembled monolithic structure can be calculated according to the specific circumstance of the structure and the value between the assembly structure and the cast-in-place structure in the table.

③ The expansion joint spacing of frame shear wall structure or frame core tube structure should be calculated according to the specific layout of the structure and the numerical value between the frame structure and the shear wall structure

④ When there is no insulation or heat insulation on the upper part of the roof, the distance between the expansion joints for the frame or shear wall structure should be taken according to the numerical value of the "open-air" in the table. The distance between the expansion joints for bent structure should be taken according to the numerical value of the "indoors or in soil" in the table.

⑤ When the column height of the frame structure (from the base top) is lower than 8m, the spacing between the expansion joints should be reduced properly

⑥ The shear wall structure of exterior wall assembly and inner wall cast-in-place wall should be used for the maximum spacing of the expansion joint according to the numerical value of the cast-in-place type. The shear wall structure constructed by sliding mode should appropriately reduce the distance between the expansion joints. In the construction of cast-in-place wall, measures should be taken to reduce the shrinkage stress of concrete.

⑦ The structure, often under high temperature, in areas of dry climate, hot summer, and frequent rainstorms, should reduce the distance between the expansion joints according to experience.

⑧ The distance between expansion joints should also consider the influence of construction conditions. If necessary, such as large shrinkage of materials or longer exposure time of indoor structures, it is advisable to reduce the space between expansion joints.

⑨ For exposed structure such as cast-in-place cornice and rain cover, the temperature expansion joint should be arranged in the longitudinal direction. Its spacing should be larger than 12m.

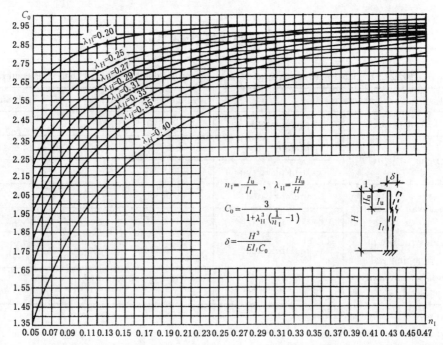

FIGURE C.1 The value of coefficient C_0 under the unit concentrated load action on the top of column

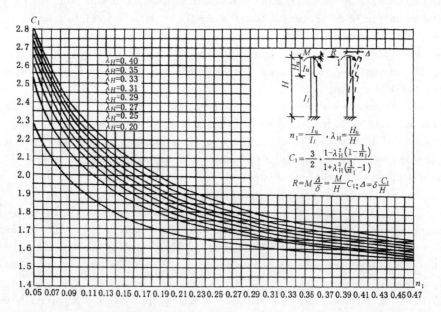

FIGURE C.2 The value of coefficient C_1 under the moment action on the top of column

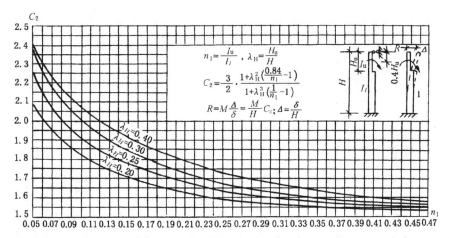

FIGURE C. 3 The value of coefficient C_2 under the moment action on the upper column ($y=0.4H_u$)

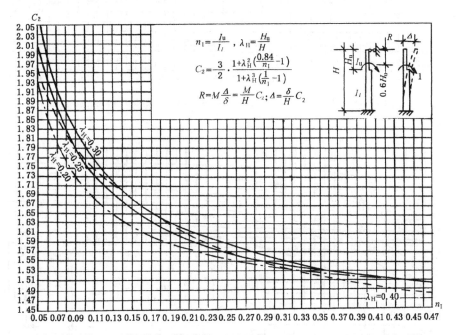

FIGURE C. 4 The value of coefficient C_2 under the moment action on the upper column ($y=0.6H_u$)

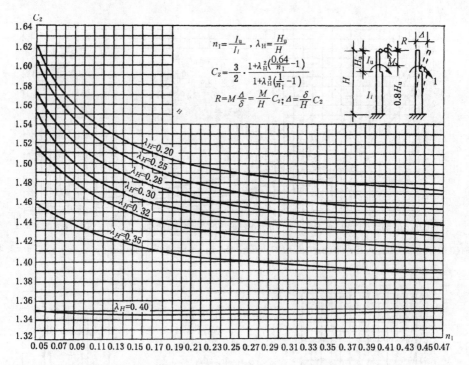

FIGURE C.5 The value of coefficient C_2 under the moment action on the upper column ($y = 0.8H_u$)

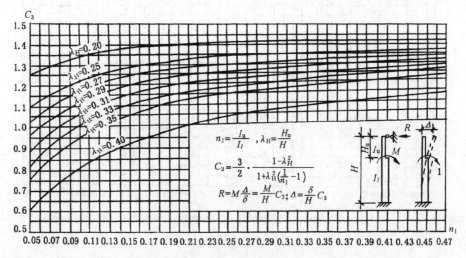

FIGURE C.6 The value of coefficient C_3 under the moment action on the corbel surface

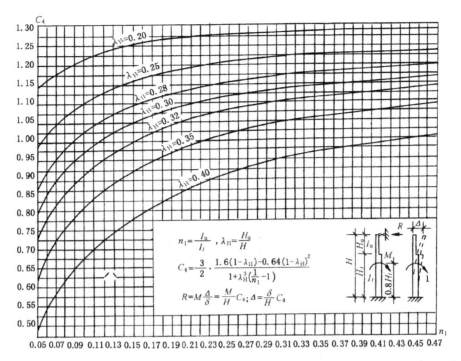

FIGURE C.7 The value of coefficient C_4 under the moment action on the lower column ($y = 0.8H_u$)

Appendix D

TABLE D.1 Standard counter bending point height ratio y_0 when the regular frame is subjected to uniform horizontal force

Total number of layers m	Layer number n	K													
		0.1	0.2	0.3	0.4	0.5	0.6	0.7	0.8	0.9	1.0	2.0	3.0	4.0	5.0
1	1	0.80	0.75	0.70	0.65	0.65	0.60	0.60	0.60	0.60	0.55	0.55	0.55	0.55	0.55
2	2	0.45	0.40	0.35	0.35	0.35	0.35	0.40	0.40	0.40	0.40	0.45	0.45	0.45	0.45
2	1	0.95	0.80	0.75	0.70	0.65	0.65	0.65	0.60	0.60	0.60	0.55	0.55	0.55	0.50
3	3	0.15	0.20	0.20	0.25	0.30	0.30	0.30	0.35	0.35	0.35	0.40	0.45	0.45	0.45
3	2	0.55	0.50	0.45	0.45	0.45	0.45	0.45	0.45	0.45	0.45	0.45	0.50	0.50	0.50
3	1	1.00	0.85	0.80	0.75	0.70	0.70	0.65	0.65	0.65	0.60	0.55	0.55	0.55	0.55
4	4	−0.05	0.05	0.15	0.20	0.25	0.30	0.30	0.35	0.35	0.35	0.40	0.45	0.45	0.45
4	3	0.25	0.30	0.30	0.35	0.35	0.40	0.40	0.40	0.40	0.45	0.45	0.50	0.50	0.50
4	2	0.60	0.55	0.50	0.50	0.45	0.45	0.45	0.45	0.45	0.45	0.50	0.50	0.50	0.50
4	1	1.10	0.90	0.80	0.75	0.70	0.70	0.65	0.65	0.65	0.60	0.55	0.55	0.55	0.55
5	5	−0.20	0.00	0.15	0.20	0.25	0.30	0.30	0.30	0.35	0.35	0.40	0.45	0.45	0.45
5	4	0.10	0.20	0.25	0.30	0.35	0.35	0.40	0.40	0.40	0.40	0.45	0.45	0.50	0.50
5	3	0.40	0.40	0.40	0.40	0.40	0.45	0.45	0.45	0.45	0.45	0.50	0.50	0.50	0.50
5	2	0.65	0.55	0.50	0.50	0.50	0.50	0.50	0.50	0.50	0.50	0.50	0.50	0.50	0.50
5	1	1.20	0.95	0.80	0.75	0.75	0.70	0.70	0.65	0.65	0.65	0.55	0.55	0.55	0.55
6	6	−0.30	0.00	0.10	0.20	0.25	0.25	0.30	0.30	0.35	0.35	0.40	0.45	0.45	0.45
6	5	0.00	0.20	0.25	0.30	0.35	0.35	0.40	0.40	0.40	0.40	0.45	0.45	0.50	0.50
6	4	0.20	0.30	0.35	0.35	0.40	0.40	0.40	0.45	0.45	0.45	0.45	0.50	0.50	0.50
6	3	0.40	0.40	0.40	0.45	0.45	0.45	0.45	0.45	0.45	0.45	0.50	0.50	0.50	0.50
6	2	0.70	0.60	0.55	0.50	0.50	0.50	0.50	0.50	0.50	0.50	0.50	0.50	0.50	0.50
6	1	1.20	0.95	0.85	0.80	0.75	0.70	0.70	0.65	0.65	0.65	0.55	0.55	0.55	0.55
7	7	−0.35	−0.05	0.10	0.20	0.20	0.25	0.30	0.30	0.35	0.35	0.40	0.45	0.45	0.45
7	6	−0.10	0.15	0.25	0.30	0.35	0.35	0.35	0.40	0.40	0.40	0.45	0.45	0.50	0.50
7	5	0.10	0.25	0.30	0.35	0.40	0.40	0.40	0.45	0.45	0.45	0.45	0.50	0.50	0.50
7	4	0.30	0.35	0.40	0.40	0.40	0.45	0.45	0.45	0.45	0.45	0.50	0.50	0.50	0.50
7	3	0.50	0.45	0.45	0.45	0.45	0.45	0.45	0.45	0.45	0.45	0.50	0.50	0.50	0.50
7	2	0.75	0.60	0.55	0.50	0.50	0.50	0.50	0.50	0.50	0.50	0.50	0.50	0.50	0.50
7	1	1.20	0.95	0.85	0.80	0.75	0.70	0.70	0.65	0.65	0.65	0.55	0.55	0.55	0.55

Appendix

Continued

Total number of layers m	Layer number n	K													
		0.1	0.2	0.3	0.4	0.5	0.6	0.7	0.8	0.9	1.0	2.0	3.0	4.0	5.0
8	8	−0.35	−0.05	0.10	0.15	0.25	0.25	0.30	0.30	0.35	0.35	0.40	0.45	0.45	0.45
	7	−1.00	0.15	0.25	0.30	0.35	0.35	0.40	0.40	0.40	0.40	0.45	0.50	0.50	0.50
	6	0.05	0.25	0.30	0.35	0.40	0.40	0.40	0.45	0.45	0.45	0.45	0.50	0.50	0.50
	5	0.20	0.30	0.35	0.40	0.40	0.40	0.45	0.45	0.45	0.45	0.50	0.50	0.50	0.50
	4	0.35	0.40	0.40	0.45	0.45	0.45	0.45	0.45	0.45	0.45	0.50	0.50	0.50	0.50
	3	0.50	0.45	0.45	0.45	0.45	0.45	0.45	0.45	0.50	0.50	0.50	0.50	0.50	0.50
	2	0.75	0.60	0.55	0.55	0.55	0.50	0.50	0.50	0.50	0.50	0.50	0.50	0.50	0.50
	1	1.20	1.00	0.85	0.80	0.80	0.75	0.70	0.65	0.65	0.65	0.55	0.55	0.55	0.55
9	9	−0.40	−0.05	0.10	0.20	0.25	0.25	0.30	0.30	0.35	0.35	0.45	0.45	0.45	0.45
	8	−0.15	1.05	0.25	0.30	0.35	0.35	0.35	0.40	0.40	0.40	0.45	0.45	0.50	0.45
	7	0.05	0.25	0.30	0.35	0.40	0.40	0.40	0.45	0.45	0.45	0.45	0.50	0.50	0.50
	6	0.15	0.30	0.35	0.40	0.40	0.45	0.45	0.45	0.45	0.45	0.50	0.50	0.50	0.50
	5	0.25	0.35	0.40	0.40	0.45	0.45	0.45	0.45	0.45	0.45	0.50	0.50	0.50	0.50
	4	0.40	0.40	0.40	0.45	0.45	0.45	0.45	0.45	0.45	0.45	0.50	0.50	0.50	0.50
	3	0.55	0.45	0.45	0.45	0.45	0.45	0.45	0.45	0.50	0.50	0.50	0.50	0.50	0.50
	2	0.80	0.65	0.55	0.55	0.50	0.50	0.50	0.50	0.50	0.50	0.50	0.50	0.50	0.50
	1	1.20	1.00	0.85	0.80	0.75	0.70	0.70	0.65	0.65	0.65	0.55	0.55	0.55	0.55
10	10	−0.40	−0.05	0.10	0.20	0.25	0.30	0.30	0.30	0.35	0.40	0.40	0.45	0.45	0.45
	9	−0.15	0.15	0.25	0.30	0.35	0.35	0.40	0.40	0.40	0.45	0.45	0.45	0.50	0.50
	8	0.00	0.25	0.30	0.35	0.40	0.40	0.40	0.45	0.45	0.45	0.45	0.50	0.50	0.50
	7	0.10	0.30	0.35	0.40	0.40	0.45	0.45	0.45	0.45	0.50	0.50	0.50	0.50	0.50
	6	0.20	0.35	0.40	0.40	0.45	0.45	0.45	0.45	0.45	0.50	0.50	0.50	0.50	0.50
	5	0.30	0.40	0.40	0.45	0.45	0.45	0.45	0.45	0.45	0.50	0.50	0.50	0.50	0.50
	4	0.40	0.40	0.45	0.45	0.45	0.45	0.45	0.45	0.45	0.50	0.50	0.50	0.50	0.50
	3	0.55	0.50	0.45	0.45	0.45	0.50	0.50	0.50	0.50	0.50	0.50	0.50	0.50	0.50
	2	0.80	0.65	0.55	0.55	0.55	0.50	0.50	0.50	0.50	0.50	0.50	0.50	0.50	0.50
	1	1.30	1.00	0.85	0.80	0.75	0.70	0.70	0.65	0.65	0.60	0.60	0.55	0.55	0.55
11	11	−0.40	−0.05	−0.10	0.20	0.25	0.30	0.30	0.30	0.35	0.35	0.40	0.45	0.45	0.45
	10	−0.15	0.15	0.25	0.30	0.35	0.35	0.40	0.40	0.40	0.40	0.45	0.45	0.50	0.50
	9	0.00	0.25	0.30	0.35	0.40	0.40	0.40	0.45	0.45	0.45	0.45	0.50	0.50	0.50
	8	0.10	0.30	0.35	0.40	0.40	0.45	0.45	0.45	0.45	0.45	0.50	0.50	0.50	0.50
	7	0.20	0.35	0.40	0.45	0.45	0.45	0.45	0.45	0.45	0.45	0.50	0.50	0.50	0.50
	6	0.25	0.35	0.40	0.45	0.45	0.45	0.45	0.45	0.45	0.45	0.50	0.50	0.50	0.50
	5	0.35	0.40	0.40	0.45	0.45	0.45	0.45	0.45	0.45	0.45	0.50	0.50	0.50	0.50
	4	0.40	0.45	0.45	0.45	0.45	0.45	0.45	0.50	0.50	0.50	0.50	0.50	0.50	0.50

Continued

Total number of layers m	Layer number n	K													
		0.1	0.2	0.3	0.4	0.5	0.6	0.7	0.8	0.9	1.0	2.0	3.0	4.0	5.0
11	3	0.55	0.50	0.50	0.50	0.50	0.50	0.50	0.50	0.50	0.50	0.50	0.50	0.50	0.50
	2	0.80	0.65	0.60	0.55	0.55	0.50	0.50	0.50	0.50	0.50	0.50	0.50	0.50	0.50
	1	1.30	1.00	0.85	0.80	0.75	0.70	0.70	0.65	0.65	0.65	0.60	0.55	0.55	0.55
Over 12	↓1	−0.40	−0.05	0.10	0.20	0.25	0.30	0.30	0.30	0.35	0.35	0.40	0.45	0.45	0.45
	2	−0.15	0.15	0.25	0.30	0.35	0.35	0.40	0.40	0.40	0.40	0.45	0.45	0.50	0.50
	3	0.00	0.25	0.30	0.35	0.40	0.40	0.40	0.45	0.45	0.45	0.50	0.50	0.50	0.50
	4	0.10	0.30	0.35	0.40	0.40	0.45	0.45	0.45	0.45	0.45	0.50	0.50	0.50	0.50
	5	0.20	0.35	0.45	0.40	0.45	0.45	0.45	0.45	0.45	0.45	0.50	0.50	0.50	0.50
	6	0.25	0.35	0.40	0.45	0.45	0.45	0.45	0.45	0.45	0.45	0.50	0.50	0.50	0.50
	7	0.30	0.40	0.40	0.45	0.45	0.45	0.45	0.45	0.50	0.50	0.50	0.50	0.50	0.50
	8	0.35	0.40	0.45	0.45	0.45	0.45	0.45	0.50	0.50	0.50	0.50	0.50	0.50	0.50
	Middle	0.40	0.40	0.45	0.45	0.45	0.45	0.50	0.50	0.50	0.50	0.50	0.50	0.50	0.50
	4	0.45	0.45	0.45	0.45	0.50	0.50	0.50	0.50	0.50	0.50	0.50	0.50	0.50	0.50
	3	0.60	0.50	0.50	0.50	0.50	0.50	0.50	0.50	0.50	0.50	0.50	0.50	0.50	0.50
	2	0.80	0.65	0.60	0.55	0.55	0.50	0.50	0.50	0.50	0.50	0.50	0.50	0.50	0.50
	↑1	1.30	1.00	0.85	0.80	0.75	0.70	0.70	0.65	0.65	0.65	0.55	0.55	0.55	0.55

TABLE D.2 Standard counter bend point height ratio y_0 when the regular frame is subjected to the horizontal force of the inverted triangle

Total number of layers m	Layer number n	K													
		0.1	0.2	0.3	0.4	0.5	0.6	0.7	0.8	0.9	1.0	2.0	3.0	4.0	5.0
1	1	0.80	0.75	0.70	0.65	0.65	0.60	0.60	0.60	0.60	0.55	0.55	0.55	0.55	0.55
2	2	0.50	0.45	0.40	0.40	0.40	0.40	0.40	0.40	0.40	0.45	0.45	0.45	0.45	0.50
	1	1.00	0.85	0.75	0.70	0.70	0.65	0.65	0.65	0.60	0.60	0.55	0.55	0.55	0.55
3	3	0.25	0.25	0.25	0.30	0.30	0.35	0.35	0.35	0.40	0.40	0.45	0.45	0.45	0.50
	2	0.60	0.50	0.50	0.50	0.50	0.45	0.45	0.45	0.45	0.45	0.50	0.50	0.50	0.50
	1	1.15	0.90	0.80	0.75	0.75	0.70	0.70	0.65	0.65	0.65	0.60	0.55	0.55	0.55
4	4	0.10	0.15	0.20	0.25	0.30	0.30	0.35	0.35	0.35	0.40	0.45	0.45	0.45	0.45
	3	0.35	0.35	0.35	0.40	0.40	0.40	0.40	0.45	0.45	0.45	0.45	0.50	0.50	0.50
	2	0.70	0.60	0.55	0.50	0.50	0.50	0.50	0.50	0.50	0.50	0.50	0.50	0.50	0.50
	1	1.20	0.95	0.85	0.80	0.75	0.70	0.70	0.70	0.65	0.65	0.55	0.55	0.55	0.55
5	5	−0.05	0.10	0.20	0.25	0.30	0.30	0.35	0.35	0.35	0.35	0.40	0.45	0.45	0.45
	4	0.20	0.25	0.35	0.35	0.40	0.40	0.40	0.40	0.40	0.45	0.45	0.50	0.50	0.50
	3	0.45	0.40	0.45	0.45	0.45	0.45	0.45	0.45	0.45	0.45	0.50	0.50	0.50	0.50
	2	0.75	0.60	0.55	0.55	0.50	0.50	0.50	0.50	0.50	0.50	0.50	0.50	0.50	0.50
	1	1.30	1.00	0.85	0.80	0.75	0.70	0.70	0.65	0.65	0.65	0.65	0.55	0.55	0.55
6	6	−0.15	0.05	0.15	0.20	0.25	0.30	0.30	0.35	0.35	0.35	0.40	0.45	0.45	0.45
	5	0.10	0.25	0.30	0.35	0.35	0.40	0.40	0.40	0.45	0.45	0.45	0.50	0.50	0.50
	4	0.30	0.35	0.40	0.40	0.45	0.45	0.45	0.45	0.45	0.45	0.50	0.50	0.50	0.50
	3	0.50	0.45	0.45	0.45	0.45	0.45	0.45	0.45	0.45	0.50	0.50	0.50	0.50	0.50
	2	0.80	0.65	0.55	0.55	0.55	0.50	0.50	0.50	0.50	0.50	0.50	0.50	0.50	0.50
	1	1.30	1.00	0.85	0.80	0.75	0.70	0.70	0.65	0.65	0.65	0.60	0.55	0.55	0.55
7	7	−0.20	0.05	0.15	0.20	0.25	0.30	0.30	0.35	0.35	0.35	0.45	0.45	0.45	0.45
	6	0.05	0.20	0.30	0.35	0.35	0.40	0.40	0.40	0.40	0.45	0.45	0.50	0.50	0.50
	5	0.20	0.30	0.35	0.40	0.40	0.45	0.45	0.45	0.45	0.45	0.50	0.50	0.50	0.50
	4	0.35	0.40	0.40	0.45	0.45	0.45	0.45	0.45	0.45	0.45	0.50	0.50	0.50	0.50
	3	0.55	0.50	0.50	0.50	0.50	0.50	0.50	0.50	0.50	0.50	0.50	0.50	0.50	0.50
	2	0.80	0.65	0.60	0.55	0.55	0.55	0.50	0.50	0.50	0.50	0.50	0.50	0.50	0.50
	1	1.30	1.00	0.90	0.80	0.75	0.70	0.70	0.70	0.65	0.65	0.60	0.55	0.55	0.55
8	8	−0.20	0.05	0.15	0.20	0.25	0.30	0.30	0.30	0.35	0.35	0.45	0.45	0.45	0.45
	7	0.00	0.20	0.30	0.35	0.35	0.40	0.40	0.40	0.40	0.45	0.45	0.50	0.50	0.50
	6	0.15	0.30	0.35	0.40	0.40	0.45	0.45	0.45	0.45	0.45	0.50	0.50	0.50	0.50
	5	0.30	0.40	0.40	0.45	0.45	0.45	0.45	0.45	0.45	0.45	0.50	0.50	0.50	0.50
	4	0.40	0.45	0.45	0.45	0.45	0.45	0.45	0.45	0.50	0.50	0.50	0.50	0.50	0.50
	3	0.60	0.50	0.50	0.50	0.50	0.50	0.50	0.50	0.50	0.50	0.50	0.50	0.50	0.50
	2	0.85	0.65	0.60	0.55	0.55	0.55	0.50	0.50	0.50	0.50	0.50	0.50	0.50	0.50
	1	1.30	1.00	0.90	0.80	0.75	0.70	0.70	0.70	0.70	0.65	0.60	0.55	0.55	0.55

Continued

Total number of layers m	Layer number n	K													
		0.1	0.2	0.3	0.4	0.5	0.6	0.7	0.8	0.9	1.0	2.0	3.0	4.0	5.0
9	9	−0.25	0.00	0.15	0.20	0.25	0.30	0.30	0.35	0.35	0.40	0.45	0.45	0.45	0.45
	8	0.00	0.20	0.30	0.35	0.35	0.40	0.40	0.40	0.40	0.45	0.45	0.50	0.50	0.50
	7	0.15	0.30	0.35	0.40	0.40	0.45	0.45	0.45	0.45	0.45	0.50	0.50	0.50	0.50
	6	0.25	0.35	0.40	0.40	0.45	0.45	0.45	0.45	0.45	0.50	0.50	0.50	0.50	0.50
	5	0.35	0.40	0.45	0.45	0.45	0.45	0.45	0.45	0.50	0.50	0.50	0.50	0.50	0.50
	4	0.45	0.45	0.45	0.45	0.45	0.50	0.50	0.50	0.50	0.50	0.50	0.50	0.50	0.50
	3	0.60	0.50	0.50	0.50	0.50	0.50	0.50	0.50	0.50	0.50	0.50	0.50	0.50	0.50
	2	0.85	0.65	0.60	0.55	0.55	0.55	0.55	0.50	0.50	0.50	0.50	0.50	0.50	0.50
	1	1.35	1.00	0.90	0.80	0.75	0.75	0.70	0.70	0.65	0.65	0.60	0.55	0.55	0.55
10	10	−0.25	0.00	0.15	0.20	0.25	0.30	0.30	0.35	0.35	0.40	0.45	0.45	0.45	0.45
	9	−0.10	0.20	0.30	0.35	0.35	0.40	0.40	0.40	0.40	0.45	0.45	0.50	0.50	0.50
	8	0.10	0.30	0.35	0.40	0.40	0.40	0.45	0.45	0.45	0.45	0.50	0.50	0.50	0.50
	7	0.20	0.35	0.40	0.40	0.45	0.45	0.45	0.45	0.45	0.50	0.50	0.50	0.50	0.50
	6	0.30	0.40	0.40	0.45	0.45	0.45	0.45	0.45	0.45	0.50	0.50	0.50	0.50	0.50
	5	0.40	0.45	0.45	0.45	0.45	0.45	0.45	0.50	0.50	0.50	0.50	0.50	0.50	0.50
	4	0.50	0.45	0.45	0.45	0.50	0.50	0.50	0.50	0.50	0.50	0.50	0.50	0.50	0.50
	3	0.60	0.55	0.50	0.50	0.50	0.50	0.50	0.50	0.50	0.50	0.50	0.50	0.50	0.50
	2	0.85	0.65	0.60	0.55	0.55	0.55	0.55	0.50	0.50	0.50	0.50	0.50	0.50	0.50
	1	1.35	1.00	0.90	0.80	0.75	0.75	0.70	0.70	0.65	0.65	0.60	0.55	0.55	0.55
11	11	−0.25	0.00	0.15	0.20	0.25	0.30	0.30	0.30	0.35	0.35	0.45	0.45	0.45	0.45
	10	−0.05	0.20	0.25	0.30	0.35	0.40	0.40	0.40	0.40	0.45	0.45	0.50	0.50	0.50
	9	0.10	0.30	0.35	0.40	0.40	0.40	0.45	0.45	0.45	0.45	0.50	0.50	0.50	0.50
	8	0.20	0.35	0.40	0.40	0.45	0.45	0.45	0.45	0.45	0.50	0.50	0.50	0.50	0.50
	7	0.25	0.40	0.40	0.45	0.45	0.45	0.45	0.45	0.45	0.50	0.50	0.50	0.50	0.50
	6	0.35	0.40	0.40	0.45	0.45	0.45	0.45	0.50	0.50	0.50	0.50	0.50	0.50	0.50
	5	0.40	0.45	0.45	0.45	0.45	0.50	0.50	0.50	0.50	0.50	0.50	0.50	0.50	0.50
	4	0.50	0.50	0.50	0.50	0.50	0.50	0.50	0.50	0.50	0.50	0.50	0.50	0.50	0.50
	3	0.65	0.55	0.60	0.50	0.50	0.50	0.50	0.50	0.50	0.50	0.50	0.50	0.50	0.50
	2	0.85	0.65	0.60	0.55	0.55	0.55	0.55	0.50	0.50	0.50	0.50	0.50	0.50	0.50
	1	1.35	1.05	0.90	0.80	0.75	0.75	0.70	0.70	0.65	0.65	0.60	0.55	0.55	0.55
Over 12	↓ From the top floor 1	−0.30	0.00	0.15	0.20	0.25	0.30	0.30	0.30	0.35	0.35	0.40	0.45	0.45	0.45
	2	−0.10	0.20	0.25	0.30	0.35	0.40	0.40	0.40	0.40	0.40	0.45	0.45	0.45	0.50
	3	0.05	0.25	0.35	0.40	0.40	0.40	0.45	0.45	0.45	0.45	0.45	0.50	0.50	0.50
	4	0.15	0.30	0.40	0.40	0.45	0.45	0.45	0.45	0.45	0.45	0.45	0.50	0.50	0.50

Continued

Total number of layers m	Layer number n	K													
		0.1	0.2	0.3	0.4	0.5	0.6	0.7	0.8	0.9	1.0	2.0	3.0	4.0	5.0
Over 12	5	0.25	0.35	0.50	0.45	0.45	0.45	0.45	0.45	0.45	0.45	0.50	0.50	0.50	0.50
	6	0.30	0.40	0.50	0.45	0.45	0.45	0.45	0.50	0.45	0.50	0.50	0.50	0.50	0.50
	7	0.35	0.40	0.55	0.45	0.45	0.45	0.50	0.50	0.50	0.50	0.50	0.50	0.50	0.50
	8	0.35	0.45	0.55	0.45	0.50	0.50	0.50	0.50	0.50	0.50	0.50	0.50	0.50	0.50
	Middle	0.45	0.45	0.55	0.45	0.50	0.50	0.50	0.50	0.50	0.50	0.50	0.50	0.50	0.50
	4	0.55	0.50	0.50	0.50	0.50	0.50	0.50	0.50	0.50	0.50	0.50	0.50	0.50	0.50
	3	0.65	0.55	0.50	0.50	0.50	0.50	0.50	0.50	0.50	0.50	0.50	0.50	0.50	0.50
	2	0.70	0.70	0.60	0.55	0.55	0.55	0.55	0.50	0.50	0.50	0.50	0.50	0.50	0.50
	↑ Start from the bottom 1	1.35	1.05	0.90	0.80	0.75	0.70	0.70	0.70	0.65	0.65	0.60	0.55	0.55	0.55

TABLE D.3 The upper and lower beam stiffness changes inflection point correction of standard height ratio y_1

α_1 \ K	0.1	0.2	0.3	0.4	0.5	0.6	0.7	0.8	0.9	1.0	2.0	3.0	4.0	5.0
0.4	0.55	0.40	0.30	0.25	0.20	0.20	0.20	0.15	0.15	0.15	0.05	0.05	0.05	0.05
0.5	0.45	0.30	0.20	0.20	0.15	0.15	0.15	0.10	0.10	0.10	0.05	0.05	0.05	0.05
0.6	0.30	0.20	0.15	0.15	0.10	0.10	0.10	0.10	0.05	0.05	0.05	0	0	0
0.7	0.20	0.15	0.10	0.10	0.10	0.10	0.05	0.05	0.05	0.05	0.05	0	0	0
0.8	0.15	0.10	0.05	0.05	0.05	0.05	0.05	0.05	0.05	0	0	0	0	0
0.9	0.05	0.05	0.05	0.05	0	0	0	0	0	0	0	0	0	0

Notice: $\alpha_1=(i_1+i_2)/(i_3+i_4)$; when $(i_1+i_2)>(i_3+i_4)$, α_1 takes the equation, $\alpha_1=(i_3+i_4)/(i_1+i_2)$, and y_1 take a minus sign; The bottom column does not consider this correction, $y_1=0$.

TABLE D. 4 The correction value of the height change of upper and lower layers on the height ratio of the standard counter bend point y_2 and y_3

α_2	α_1	K=0.1	0.2	0.3	0.4	0.5	0.6	0.7	0.8	0.9	1.0	2.0	3.0	4.0	5.0
2.0		0.25	0.15	0.15	0.10	0.10	0.10	0.10	0.10	0.05	0.05	0.05	0.05	0	0
1.8		0.20	0.15	0.10	0.10	0.10	0.05	0.05	0.05	0.05	0.05	0.05	0	0	0
1.6	0.4	0.15	0.10	0.10	0.05	0.05	0.05	0.05	0.05	0.05	0.05	0	0	0	0
1.4	0.6	0.10	0.05	0.05	0.05	0.05	0.05	0.05	0.05	0.05	0	0	0	0	0
1.2	0.8	0.05	0.05	0.05	0	0	0	0	0	0	0	0	0	0	0
1.0	1.0	0	0	0	0	0	0	0	0	0	0	0	0	0	0
0.8	1.2	−0.05	−0.05	−0.05	0	0	0	0	0	0	0	0	0	0	0
0.6	1.4	−0.10	−0.05	−0.05	−0.05	−0.05	−0.05	−0.05	−0.05	0	0	0	0	0	0
0.4	1.6	−0.15	−0.10	−0.10	−0.05	−0.05	−0.05	−0.05	−0.05	−0.05	−0.05	0	0	0	0
	1.8	−0.20	−0.15	−0.10	−0.10	−0.10	−0.05	−0.05	−0.05	−0.05	−0.05	−0.05	0	0	0
	2.0	−0.25	−0.15	−0.15	−0.10	−0.10	−0.10	−0.10	−0.10	−0.05	−0.05	−0.05	−0.05	0	0

Notice: $\alpha_2 = h(\text{up})/h$, $\alpha_3 = h(\text{down})/h$, h is a computing layer number, $h(\text{up})$ is upper layer, $h(\text{down})$ the next layer of height; y_2 lookup by K and α_2. Do not consider this amendment to the top level; y_3 looks up the table by K and α_3, and does not consider this amendment to the bottom.

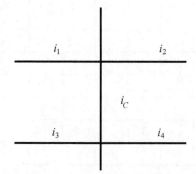